SpringerBriefs in Molecular Science

Protein Folding and Structure

Series editor

Cláudio M. Gomes, Biosystems & Integrative Sciences Institute, Faculty Sciences University Lisboa, Lisboa, Portugal

About the Series

Prepared by leading experts, the Springer Briefs subseries on *Protein Folding and Structure* contains diverse types of contributions, from snapshot volumes that allow fast entry to a general topic to those covering more specialized aspects in the field of protein folding and structure. In common, these *Briefs* aim at covering essential concepts, methodologies and ideas in the context of contemporary research in protein science. Through these compact volumes, this series serves as a venue for publication between typical research papers, review articles and full books, and aims at a broad audience, from students to researchers in academia and industry.

About the Editor

Cláudio M. Gomes is Associate Professor with aggregation at the Faculty of Sciences University of Lisboa where he heads the Protein Folding and Misfolding Laboratory as part of BioISI Biosystems and Integrative Sciences Institute. He obtained is PhD in Biochemistry (1999) from the Universidade Nova de Lisboa, as a graduate of the Gulbenkian Ph.D. program in Biology and Medicine, and holds Habilitation (Agregação) in Biochemistry (2013). He has extensive publishing and editorial activities, both as a prolific author, member of Editorial boards and editor of thematic issues and books. In collaboration with Springer, he set the Springer Briefs subseries on Protein Folding and Structure, which launched its first volume in 2014.

More information about this series at http://www.springer.com/series/11958

Carlos A. Salgueiro · Joana M. Dantas

Multiheme Cytochromes

 Springer

Carlos A. Salgueiro
UCIBIO-Requimte, Departamento de
  Química
Faculdade de Ciências e Tecnologia,
  Universidade Nova de Lisboa
Caparica
Portugal

Joana M. Dantas
UCIBIO-Requimte, Departamento de
  Química
Faculdade de Ciências e Tecnologia,
  Universidade Nova de Lisboa
Caparica
Portugal

ISSN 2191-5407   ISSN 2191-5415   (electronic)
SpringerBriefs in Molecular Science
ISSN 2199-3157   ISSN 2199-3165   (electronic)
Protein Folding and Structure
ISBN 978-3-642-44960-4   ISBN 978-3-642-44961-1   (eBook)
DOI 10.1007/978-3-642-44961-1

Library of Congress Control Number: 2016948115

Printed on acid-free paper

This Springer imprint is published by Springer Nature
The registered company is Springer-Verlag GmbH Berlin Heidelberg

Foreword

Proteins are extremely versatile biomolecules involved in a wide range of functions in the cell. Among these, redox catalysis and electron transfer processes are primordial for metabolic reactions and cellular bioenergetics. Such reactions are made possible by the association of redox cofactors to proteins, which in many cases ultimately also play important structural roles. Among these, heme plays a crucial role as its association to proteins greatly expands their functional landscapes in redox processes. A plethora of hemeproteins, which are found in different structural flavors, occasionally combining additional redox cofactors with heme, assures a vast number of biological functions. In the fourth volume of the Protein Folding and Structure series of the Springer Briefs in Molecular Science, we get a unique perspective into multiheme cytochromes. Through the insight of the leading expert Carlos A. Salgueiro, the reader is presented with an in-depth comprehensive overview of the biology, structural and functional roles of these complex heme proteins, their vital biological roles and biotechnological impacts, ranging from bioremediation to bioenergy production. Enjoy reading!

Lisboa
August 2016

Cláudio M. Gomes
Editor
Springer Briefs series on protein folding and structure

Preface

The purpose of this SpringerBriefs volume is to provide an overview of the functional roles of multiheme cytochromes and to describe the recent progresses in their structural and functional characterization. It starts by revisiting the diversity of cytochromes, providing details on their main features and biogenesis. The chemistry of cytochromes is extremely heterogeneous, which reflects their different amino acid composition, sequence and tertiary arrangement around the heme groups. Such functional heterogeneity is even more striking in the case of multiheme cytochromes, in which the presence of several heme groups favors the efficient conversion of a substrate, extends the protein's functional redox potential range, allows the cooperative transfer of electrons, enhances the bacterial electron-storage capacity and the electron transfer through large distances without the need for successive binding events. Multiheme cytochromes are abundant in dissimilatory metal reducing bacteria, particularly in the *Shewanella* and *Geobacter* genera. These microorganisms can couple the oxidation of cytoplasmic compounds to the reduction of extracellular electron acceptors, a process currently explored in different biotechnological applications. Multiheme cytochromes are key players in these extracellular electron transfer pathways and representative members from *Shewanella* and *Geobacter* spp. are selected to illustrate how these proteins can be functionally and structurally characterized. Given the emergent multiheme cytochromes-biotechnological based applications we hope that this SpringerBriefs will stimulate specialists in fields such as molecular dynamics, protein structure and bioenergetics to provide more detailed contributions on the characterization of these proteins and assist on the elucidation of the bacterial extracellular electron transfer mechanisms.

Caparica, Portugal

Carlos A. Salgueiro
Joana M. Dantas

Acknowledgments

We would like to thank the current and former members of our group. In particular, we would like to thank Dr. Leonor Morgado for her contribution to the characterization of *Geobacter sulfurreducens* multiheme cytochromes. The authors also are also deeply grateful to Dr. Marianne Schiffer, Dr. Raj Pokkuluri and Dr. Yuri Londer (National Argonne Laboratory—University of Chicago, USA), Prof. Derek Lovley (University of Massachusetts, USA), Prof. Marta Bruix (Instituto Química Física Rocasolano, CSIC, Madrid, Spain) and Prof. David L. Turner (Instituto Tecnologia Química e Biológica António Xavier, Oeiras, Portugal) for the very stimulating collaborations. Their expertise, enthusiasm and perseverance were a precious contribution for the actual state of the art on multiheme cytochromes.

All the other colleagues are cited for their contribution to our current understanding on multiheme cytochromes. Last but not least, Carlos A. Salgueiro deeply acknowledges Prof. António V. Xavier, who introduced pioneering and innovative methodologies in the characterization of multiheme cytochromes, for his invaluable academic and scientific advising. Research in the author's group is currently supported by project grant PTDC/BBB-BQB/3554/2014 from Fundação para a Ciência e Tecnologia (Portugal). Further support is provided by SFRH/BD/89701/2012 (to Joana Dantas) and UID/Multi/04378/2013 from Fundação para a Ciência e Tecnologia (Portugal).

Contents

About the Authors

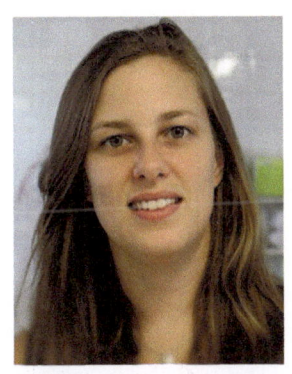

Joana M. Dantas obtained her degree in Applied Chemistry, Biotechnology from Faculty of Science and Technology, Universidade NOVA de Lisboa (FCT/UNL) in 2010. She is a member of the Biochemistry and Bioenergetics of Heme Proteins research group at UCIBIO-REQUIMTE, Chemistry Department, FCT/UNL. In 2012, she obtained her M. D. in Structural and Functional Biochemistry and she is finalizing her PhD studies. Her research is focused on the structural and dynamic characterization of multiheme cytochromes.

Carlos A. Salgueiro obtained his Ph.D. in Biochemistry from Universidade NOVA de Lisboa in 1998. Currently he is Assistant Professor with Habilitation at the Chemistry Department, FCT/UNL and Group Leader of the Biochemistry and Bioenergetics of Heme Proteins research team at UCIBIO-REQUIMTE. His research is focused on the application of NMR techniques to the structural and functional characterization of biological systems, particularly those involving electron transfer proteins. In the last decade he has been involved in the study of multiheme cytochromes from *Geobacter sulfurre-ducens*. These studies have pinpointed multiheme cytochromes as foundations to develop *Geobacter*-based applications in the fields of bioremediation and bioenergy.

About the Authors

Abbreviations

2D	Two-dimensional
ATP	Adenosine triphosphate
BET	Bioelectrochemical treatment
Ccm	Cytochrome c maturation
DMR	Dissimilatory metal reduction
DMSO	Dimethyl sulfoxide
DvHc$_3$	Tetraheme cytochrome c_3 from *Desulfovibrio vulgaris* (Hildenborough)
EET	Extracellular electron transfer
EXSY	Exchange NMR spectroscopy
HP	High potential
HSQC	Heteronuclear single-quantum coherence
IM	Inner membrane
Imc	Inner membrane cytochrome
LP	Low potential
Mac	Metal reduction associated cytochrome
MES	Microbial electrosynthesis
MFCs	Microbial fuel cells
MHC	Multiheme cytochrome c
MQ	Menaquinone
MQH$_2$	Menaquinol
MW	Molecular weight
NAD$^+$	Nicotinamide adenine dinucleotide (oxidized)
NADH	Nicotinamide adenine dinucleotide (reduced)
NMR	Nuclear magnetic resonance
NOE	Nuclear Overhauser effect
NOESY	Nuclear Overhauser spectroscopy
OM	Outer membrane
Omc	Outer membrane cytochrome
PDB	Protein Data Bank
Ppc	Periplasmic cytochrome

Sffcc$_3$	Flavocytochrome c_3 from *Shewanella frigidimarina*
SHE	Standard hydrogen electrode
SHP	Sphaeroides heme protein
Sofcc$_3$	Flavocytochrome c_3 from *Shewanella oneidensis*
STC	Small tetraheme cytochrome
TMAO	Trimethylamine oxide

Chapter 1
Multiheme Cytochromes

1.1 Cytochromes *c*

Cytochromes *c* are electron transfer proteins that participate in aerobic and anaer-obic respiratory processes and, thus, are crucial for Life. The polypeptide chain of these proteins (apoprotein) is covalently bound to at least one heme group via thioether linkages established with the sulfhydryl groups of two cysteine residues in a conserved binding motif sequence: Cys-X-X-Cys-His, where X represents any amino acid (Fig. 1.1) [1]. As depicted in Fig. 1.1, four out of the six heme iron coordination positions are equatorially occupied by the protoporphyrin IX nitrogen atoms. One of the two axial coordination positions is typically occupied by the side chain of the histidine residue in the binding motif sequence, which is also desig-nated as proximal ligand. On the other hand, the distal ligand is more variable and can be the side chain of a (i) methionine, which predominates in monoheme cytochromes *c*, (ii) histidine, particularly in multiheme cytochromes or (iii) as-paragine or tyrosine, which occur less frequently [2, 3]. In each case, the axial ligating atom is different: (i) nitrogen, in the case of histidine or asparagine, (ii) sulfur in the case of methionine or cysteine, and (iii) oxygen in the case of tyrosine. The distal position of the heme can be also transiently vacant, as observed in cytochromes with enzymatic activity [2, 3].

Heme proteins that are involved in electron transfer reactions usually have both axial positions occupied. This is the case of cytochrome PccH from *Geobacter sulfurreducens* and cytochrome c_3 from *Desulfovibrio vulgaris* (Hildenborough). PccH is a monoheme cytochrome *c* axially coordinated by histidine and methionine residues (Fig. 1.2a), whereas cytochrome c_3 is a tetraheme cytochrome *c* with all hemes axially coordinated by two histidine residues (Fig. 1.2b). On the other hand, in the case of cytochromes showing catalytic activity, the heme group is typically either five-coordinated or binds to a water molecule in the sixth coordination site in the resting state. Typically these proteins have a substrate(s) binding pocket accessible to the heme iron. This feature is exemplified by the Sphaeroides Heme

© The Author(s) 2016
C.A. Salgueiro and J.M. Dantas, *Multiheme Cytochromes*,
Protein Folding and Structure, DOI 10.1007/978-3-642-44961-1_1

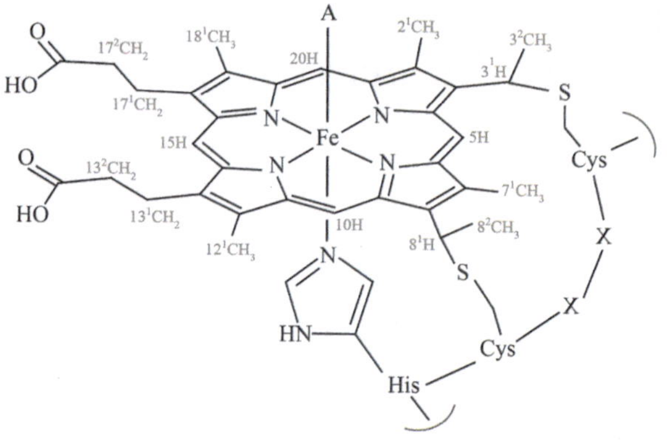

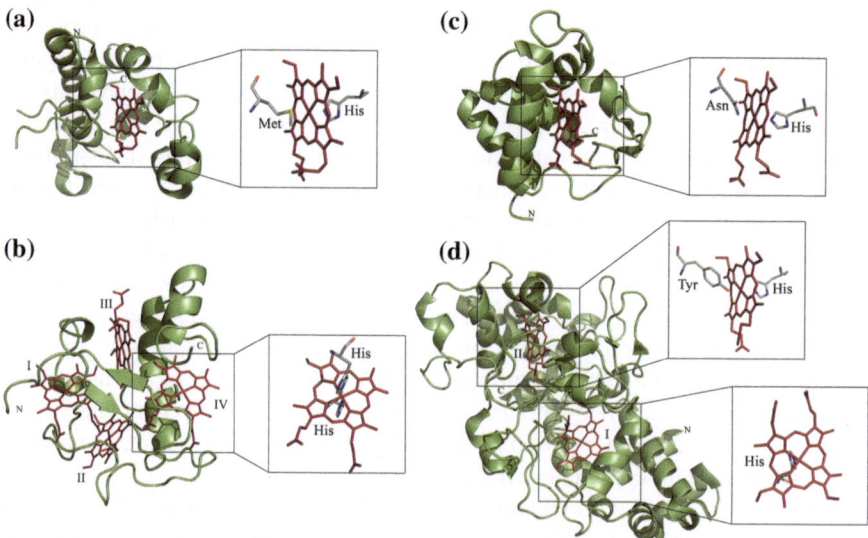

Fig. 1.1 Schematic representation of a *c*-type heme and the correspondent polypeptide binding motif. The axial coordination position labeled with A can be free or occupied by the side chain of a methionine, histidine, asparagine or tyrosine residues. The IUPAC nomenclature for tetrapyrroles is illustrated in *gray* [152]

Fig. 1.2 Structures of different cytochromes *c* showing distinct heme axial ligand coordination. Heme groups, polypeptide chain and axial ligands are colored *red*, *green* and *gray* by heteroatom, respectively. The N- and C-terminus regions are also indicated. **a** Crystal structure of monoheme cytochrome *c* PccH from *Geobacter sulfurreducens* (PDB code: 4RLR [153]); **b** Solution structure (lowest energy) of cytochrome *c*₃ from *Desulfovibrio vulgaris* (Hildenborough) (PDB code: 2CTH [154]); **c** Crystal structure of an oxygen-binding cytochrome *c* from *Rhodobacter sphaeroides* (PDB code: 1DW0 [5]); **d** Crystal structure of the diheme cytochrome *c* MauG from *Paracoccus denitrificans* (PDB code: 3L4O [155]). In each case, expansions of the heme(s) and axial ligand(s) are shown. Roman numerals indicate the hemes in their order of attachment to the Cys-X-X-Cys-His motif in the polypeptide chain. The structures were drawn using the PyMOL molecular graphics system [156]

Protein (SHP), an oxygen binding monoheme protein firstly described in purple photosynthetic bacteria [4]. The crystal structure of the SHP from *Rhodobacter sphaeroides* shows that the heme is axially coordinated by one histidine and one asparagine in the oxidized state (Fig. 1.2c). The axial asparagine residue is displaced upon reduction or binding of small molecules, such as oxygen, carbon monoxide, cyanide or nitric oxide [5]. Another example is MauG, an enzyme responsible for the post-translational modification of two tryptophan residues to form the tryptophan tryptophylquinone cofactor of methylamine dehydrogenase. MauG has two *c*-type hemes that bind one histidine as fifth axial ligand (Fig. 1.2d). The sixth axial coordination position is free in one of the hemes, whereas the other holds a tyrosine residue.

The biogenesis of cytochromes encompasses complex protein machineries named systems I–VI, which are essential for their transport and correct folding. These systems are composed by multiple proteins that assure the transport of the apoprotein and the heme group across the cytoplasmic membrane, as well as their assembling via post-translational modifications in the periplasm [6–9]. Besides differing in their apparent molecular complexity, the precise functional mechanisms of systems I-VI are still poorly understood. System I, also designated as cytochrome *c* maturation (Ccm) system is found in Gram-negative bacteria, Archaea and in the mitochondria of land plants, red algae, ciliates, jackobids. System II or cytochrome *c* synthase (Ccs) system is typically found in β-, δ-, and ε-proteobacteria, Gram-positive bacteria, Aquificales, cyanobacteria, algal and plant chloroplasts [7]. System III, also named holocytochrome *c* synthase or cytochrome *c* heme lyase, is found in the mitochondria of fungi and animals. System IV is specific for the binding of heme *c* to cytochrome b_6 in the cytochrome b_6f complex, through one thioether bond. The covalent attached heme is usually designated c_i. This system is found in all organisms with oxygenic photosynthesis but not in Firmicutes, although they also have a cytochrome *b* protein with an additional heme c_i [9]. System V is found in the mitochondria of members of the phylum Euglenozoa and, finally, system VI was identified in *Bacillus* spp. where it is responsible for the covalently heme attachment to the cytochrome *b* subunit of the cytochrome *bc* complex [8, 10].

The Ccm system is the most complex one and encompasses an ensemble of eight proteins (CcmABCDEFGH) that are responsible for the maturation of cytochromes *c* in the periplasm. With the exception of CcmA protein, which is located in the cytoplasm, all the others are associated with the cytoplasmic membrane or have periplasmic orientated domains [10]. As mentioned above, the precise cytochrome *c* maturation is not yet fully understood. However, it was shown that proteins CcmABCDE are involved in the heme handling and deliver to the apoprotein [11, 12]. The CcmF protein transfers the heme group from CcmE to the apoprotein and is also involved in the covalent binding of the two cysteines in the heme binding motif, in a process assisted by CcmG protein [13]. Finally, CcmH has a chaperon-like activity and is involved in the final heme-apocytochrome *c* ligation step [14].

1.2 Multiheme Cytochromes

Multiheme cytochromes c (MHC) are crucial components of several biological processes where they are responsible for electron transfer events. MHC can also display enzymatic activity (for a review see [15, 16] and references therein) or participate in signal transduction pathways (for a review see [17] and references therein).

In the case of MHC exhibiting catalytic activity, the presence of more than one heme is manifested. For example, the flavocytochrome c_3 from the bacterium *Shewanella frigidimarina* is a soluble and monomeric (64 kDa) periplasmic fumarate reductase [18] that catalyzes the reduction of fumarate to succinate, in a two-electron transfer event. The crystal structure of this protein reveals that it folds in three domains: a N-terminal heme domain, containing four c-types hemes each axially coordinated by two histidines; a flavin domain which contains a non-covalently bound FAD, located close to the active site; and a clamp domain [19, 20]. Therefore, the presence of more than one heme group in the vicinity of the catalytic site favors the efficient conversion of the substrate by facilitating a sequential transfer of two electrons from the heme domain to the FAD at the flavin domain.

In the case of MHC with no enzymatic activity, one evident advantage of having multiple heme groups strategically arranged along the polypeptide chain is to extend their global working redox potential range, as a result of the overlapping of the collective contribution of the individual heme redox potential ranges (Fig. 1.3). As a consequence, the functional versatility of MHC is increased, in contrast with that observed for monoheme cytochromes whose working functional ranges are limited to the Nernst response correspondent to a one-electron transfer event. Furthermore, the presence of several hemes allows the protein to receive or transfer multiple electrons in a cooperative way, a process that can be modulated by the intrinsic properties of neighboring hemes (redox interactions) or protonatable centers (redox-Bohr effect) ([21] and reference therein). Such modulation can be also be explored to optimize the electron transfer driving force or energy transduction-based biological events.

MHC can also putatively work as electron biocapacitors and contribute to the enhancement of the bacterial electron-storage capacity. In fact, Esteve-Núñez et al. [22] evaluated the electron storage capacity of *Geobacter sulfurreducens* cytochromes in absence of terminal acceptors. This study showed that cytochromes provided to the bacterium sufficient electron-storage capacity to permit a continued electron transfer across the cytoplasmic membrane to maintain energy requirements or enough proton motive force to power flagella motors involved in cellular motion.

Typically, the heme iron-iron distances between adjacent hemes in MHC do not exceed 16 Å. This allows a fast electronic exchange between the redox centers, a crucial feature to assure efficient redox reactions [3, 23]. In fact, Page et al. [24] showed that the distance between the redox centers is the most important factor in providing robustness to electron transfer in multi-redox center proteins. Therefore, depending upon the number and spatial architecture of the heme groups, MHC may allow intramolecular long-range electron transfer. A striking example is provided

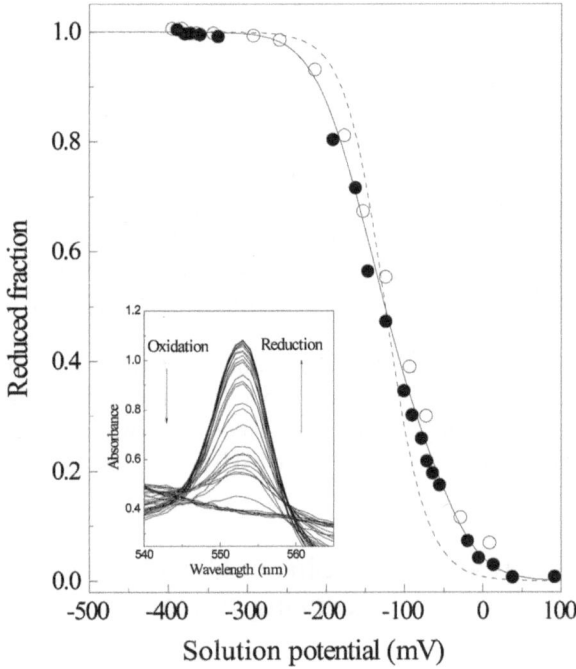

Fig. 1.3 Potentiometric redox titrations followed by visible spectroscopy of the dodecaheme cytochrome GSU1996 (pH 7 and 298 K) illustrating the wider potential range of this cytochrome when compared to a monoheme cytochrome with the same midpoint reduction potential (*dashed line*). The open and filled symbols represent the data points in the reductive and oxidative direction, respectively. *Solid lines* indicate the result of the fits for the model of consecutive reversible redox steps between the different oxidation stages. The inset shows the α-band region of the visible spectra used for the redox titration (see Sect. 1.6). This research was originally published in Dalton Transactions. Telma C. Santos, Marta A. Silva, Leonor Morgado, Joana M. Dantas and Carlos A. Salgueiro, Diving into the redox properties of *Geobacter sulfurreducens* cytochromes: a model for extracellular electron transfer. Dalton Transactions. 2015; 44: 9335-9344 © copyright holder

by three *G. sulfurreducens* cytochromes that contain twelve (GSU1996 and GSU0592; the naming scheme of the genome is according to http://www.genome.jp/kegg/genome.html) and twenty-seven (GSU2210) heme groups [25, 26]. The high number of hemes in these proteins not only provides the cell with a remarkable electron storage capacity but also allows them to transfer electrons through large distances without the need for successive binding events.

For the reasons presented above, the precise tuning of heme redox potentials is of crucial importance in MHC. These are affected by several factors that include: (i) the differences in the free energy between the oxidized and reduced states resulting from molecular interactions changes; (ii) the modulation of the electrostatic interactions within the protein or with the solvent; (iii) the heme solvent accessibility; (iv) the extent to which the heme group is distorted from planarity; (v) the protonation state of the heme propionate groups; (vi) the type of axial ligands and heme iron coordination [1, 27–40]. In most of the cases, MHC have a low amino acid residue to heme ratio compared to monoheme cytochromes and, therefore, higher heme solvent exposure [1, 30, 41]. The heme solvent exposure and axial coordination (typically His-His) strongly contribute to the low redox potential values observed for MHC [42, 43].

Detailed proteomic analysis revealed the existence of at least one MHC in members belonging to the major groups of Bacteria and Archaea, with higher incidence in Bacteria [2, 41]. Inside this domain, MHC are predominantly found in Gram-negative bacteria [2, 41]. In fact, it was observed that the distribution of MHC varies considerably among the available genomes [41]. To the best of our knowledge, the Gram-negative δ-proteobacteria from the *Geobacter*, *Shewanella*, *Anaeromyxobacter* genera, as well as the γ-proteobacteria from *Desulfovibrio* genus are the microorganisms whose genome encode for the higher number of MHC. This is particularly noticeable in *G. uraniumreducens* Rf4 whose genome encodes 75 MHC, the largest number identified so far [41].

1.3   Natural Sources of Multiheme Cytochromes: Dissimilatory Metal Reducing Bacteria

Dissimilatory metal reduction (DMR) is a widespread process that occurs in the major groups of Bacteria [44] and is fundamental for the geochemistry of aquatic sediments, submerged soils and the terrestrial subsurface [45]. Unlike other respiratory pathways, where the final electron acceptor is a freely diffusible gas or a readily soluble molecule that can be reduced inside the cells, DMR bacteria can sustain their energy requirements by coupling the oxidative metabolism of organic compounds to the reduction of extracellular metals [45, 46]. The respiratory reduction of extracellular electron acceptors can be achieved by two broad mechanisms that involve direct or indirect reduction of the acceptors [23, 47]. In the first case, direct electron transfer is achieved via (i) the contact between redox proteins located in the surface of the bacteria or (ii) electrically conductive appendages that have been called biological nanowires (Fig. 1.4). In the indirect mechanism, reduction of acceptors can be mediated by (i) small redox-active molecules that are either secreted by the cell into the environment (endogenous electron shuttles) or are already available in the environment (exogenous electron shuttles) or (ii) organic ligands produced by the bacteria, designated chelators, which solubilize metals prior to their reduction (Fig. 1.4).

In the last two decades, DMR bacteria have awakened significantly attention because of their impact on the natural environment and practical applications that include the bioremediation of organic and inorganic contaminants, bioenergy production via microbial fuel cells (MFCs) and bioelectronics. In fact, the respiratory skills of DMR bacteria provide them an important role in the environmental recycling and removal of metal contaminants from groundwaters [45]. Furthermore, some of these microorganisms (*e.g. Desulfuromonas acetoxidans, Geobacter sulfurreducens*,

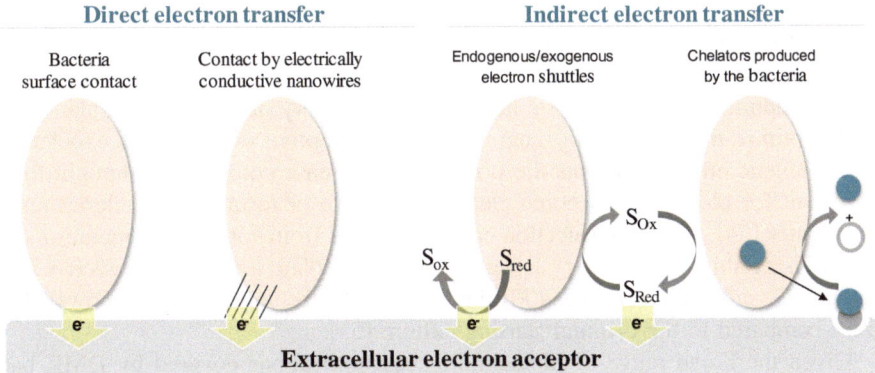

Fig. 1.4 Schematic representation of the direct and indirect electron transfer mechanisms from the bacteria to the extracellular electron acceptors. S_{ox} and S_{red} correspond to oxidized and reduced electron shuttles, respectively. The action of chelator molecules is illustrated by *circles*: *blue* (chelator molecule); *gray* (insoluble acceptor); *white* (solubilized acceptor)

Geobacter metallireducens, Rhodoferax ferrireducens, Desulfobulbus propionicus, Enterococcus gallinarum, Shewanella putrefaciens and *Shewanella oneidensis* [48]) belong to the group of exoelectrogens as they can also couple their oxidative metabolism to the electron transfer toward electrode surfaces from which electric current can be harvested in MFCs [49, 50]. Particularly in the case of *Geobacter* spp., highly cohesive protein filaments with metallic-like conductivity can be produced, offering the possibility to design microbial-based electronic sensors and other devices [51].

Currently much research is focused on alternative and renewable energy sources to offset world-wide concerns regarding global warming and fossil energy depletion [52]. Understanding the electron transfer processes by which exoelectrogens can couple the oxidation of organic compounds to the transfer of electrons to electrodes is a crucial step envisaging the optimization of current production by MFCs. Although the present MFC power densities are too low to be considered as a viable alternative energy source, substantial improvements have been made in the last years (for a review see Ren et al. [53] and references therein). In a MFC, the anode and cathode compartments are often separated by a proton selective membrane. The bacteria on the anode surface oxidize organic compounds to carbon dioxide with concomitant transfer of electrons to the anode. Electrons are then sent to the cathode via an electric circuit holding a resistor or other types of electric devices. The protons released at the anode migrate to the cathode compartment across the proton selective membrane where they combine with electrons and oxygen to form water [50]. The current low power density of MFC is primarily attributed to the ineffective electron transfer from microbes located far from the anode and proton accumulation at the anode compartment that suppresses microbial activity. Abiotic approaches have been implemented to increase the current density in MFCs including (i) the improvement of the reaction rates at the cathode, (ii) the improvement of mass transfer of organic compounds/H^+ carrying buffers; (iii) the implementation of electrode materials with higher surface area to volume ratios and electrical conductivities; (iv) the improvement of MFCs configuration or (v) the reduction of electrode resistances (for a review see Franks et al. [52]; Ren et al. [53]). In addition to these abiotic factors, there is plenty of room for cellular-based MFC optimization by genetic engineering and adaptive evolution of exoelectrogens. Genetic engineering has the potential to increase components thought to be important for biofilms to become electrically active or to increase bacterial respiration rates [54]. Adaptive selection can be obtained from continuous operation of a MFC. Using this methodology it was possible to isolate a *G. sulfurreducens* DL1 variant (*G. sulfurreducens* KN400) that produced substantially higher power densities compared to the original starting culture [55].

Given the broad range of biotechnological applications covered by DMR bacteria it is important to understand the functional mechanism of their respiratory pathways. This would allow the scientific community to improve and optimize DMR-based practical applications. Of these, the most studied DMR bacteria belong to the *Shewanellaceae* and *Geobacteraceae* families, in particularly *S. oneidensis* and *G. sulfurreducens*. These bacteria display a remarkably respiratory versatility, a hallmark underlying their proliferation in the environment. Both bacteria contain high number of genes encoding for cytochromes *c* thought to be involved in different electron transfer pathways. The analysis of the genome sequence of *G. sulfurreducens* [56] revealed that this bacterium encodes 111 cytochromes *c*, 73 of which are MHC. In the case of *S. oneidensis* [57] its genome contains at least 39 genes encoding for cytochromes *c*, 14 of which corresponding to MHC. Gene knock-out studies have shown that deletion of *G. sulfurreducens* or *S. oneidensis* MHC affects the current production and/or the reduction of extracellular acceptors by these bacteria [23, 58–62]. The key role of these proteins in extracellular electron transfer (EET) was also supported by proteomic studies that showed increased expression of MHC in the presence of such electron acceptors. The strategic localization of MHC at the bacterial inner membrane (IM), periplasm or outer membrane (OM) allows the transfer of electrons from intracellular carriers (*e.g.* NADH) to extracellular acceptors.

1.4   Recombinant Production of Multiheme Cytochromes

Given the biological importance of MHC, their structural and functional characterization is of upmost importance to elucidate EET mechanisms and establish foundations to improve EET-based practical applications. In order to achieve these goals, considerable amounts of protein are necessary. However, the natural sources of MHC are, in general, anaerobic bacteria whose cultures are difficult to manipulate and make the isolation of MHC a lengthy and expensive process. Therefore, several approaches have been developed to improve the recombinant production of MHC, a complex process that requires the correct attachment of each heme to the apoprotein and the precise axial coordination of the heme iron.

Escherichia coli is a Gram-negative and facultative anaerobic bacterium commonly used as the first choice to express recombinant proteins. The genetic systems developed for *E. coli* and the availability of large number of cloning vectors makes it the preferred microorganism for the aerobic production of heterologous proteins. As mentioned in Sect. 1.1, the maturation of cytochromes *c* involves the enzymatic complex CcmABCDEFGH formed by eight proteins. Even though the *ccm* genes are present on the *E. coli* chromosome, they are not expressed under aerobic growth conditions. Several attempts to express genes encoding MHC in *E. coli* have failed, resulting in recombinant proteins that either accumulated as precursors in the cytoplasmic membrane [63] or are exported to the periplasm as apoproteins [64]. Only at the end of the 20th century, it became possible to increase the efficiency of the post-translational maturation of MHC by co-expressing the plasmid encoding the desired cytochrome together with a second plasmid, pEC86, that contains the *ccmABCDEFGH* gene cluster [65–71].

Since then, several efforts have been undertaken to increase the MHC expression yields. These included the optimization of the expression constructs as well as the host strains [72, 73]. In 2002, Londer et al. [72] developed an expression system to overproduce *G. sulfurreducens* MHC in *E. coli*. In the particular case of triheme cytochromes, the encoding gene was cloned in a pASK40-based construct containing the *lac* promoter [72]. The co-expression of the pASK40-based construct and pEC86 increased the protein expression yields from 1 mg/L or less, previously obtained with different methodologies, up to 6 mg/L of culture of correctly folded recombinant cytochrome.

An alternative approach to the recombinant production of MHC was carried out using *S. oneidensis* cells [74, 75]. This method takes advantage of the host cell's own mechanism for heme insertion. However, in its original version the protocol used a ligase-based method for gene cloning making challenging its implementation for high-throughput cloning and expression. Shi et al. [73] extended this methodology by developing a system suitable for high level expression and purification of MHC. In this case, the expression vector pBAD202/D-TOPO, which is a pUC-like plasmid that replicates in *S. oneidensis* cells, leading to an increase of the expression levels from 0.01 to 0.9 % of total soluble proteins [73]. This methodology was first demonstrated for MtrA, a decaheme cytochrome *c* from *S. oneidensis* and DVU3171, a tetraheme cytochrome c_3 from *D. vulgaris* (Hildenborough). This procedure allowed the efficient homologous or heterologous overexpression of MHC in *S. oneidensis*. It is worth noting that the genes cloned into pBAD202/D-TOPO also incorporate a His_6 tag at the C-terminal of the expressed proteins. The use of histidine tags to facilitate subsequent protein detection and isolation requires carefulness, because the MHC heme groups are axially coordinated by two histidine residues in most of the cases. The recombinant expression of MHC cytochromes from *Shewanella* and *Geobacter*, together with genetic, proteomic and biochemical studies, shed light on the present understanding of the electron transfer pathways in these bacteria.

1.5 Extracellular Electron Transfer Pathways Involving Multiheme Cytochromes

1.5.1 Extracellular Electron Transfer Pathways in Shewanella spp.

Bacteria from the genus *Shewanella* are Gram-negative facultative anaerobes found in marine and freshwater environments [76]. These bacteria are able to use a diversity of terminal electron acceptors in anoxic environments, such as fumarate, nitrate, nitrite, trimethylamine oxide (TMAO), dimethyl sulfoxide (DMSO), sulfite, thiosulfate, elemental sulfur and metals, such as Fe(III), Mn(III), Mn(IV) and U(VI) that may be present either as soluble complexes or solid mineral (hydr)oxides [23]. This respiratory capability can be used for several practical applications. For example, at physiological conditions, the reduction of soluble metals (*e.g.* U(VI), Cr (VI)) to insoluble precipitates (U(IV), Cr(III), respectively) facilitates the bioremediation of these pollutants. Additionally, these bacteria can be used to develop bioenergy applications as they grow on several electrode surfaces [77].

Two EET mechanisms have been proposed for *S. oneidensis*: (i) direct electron transfer, by the contact between the extracellular acceptor either with the surface of the bacteria or with the nanowires; (ii) indirect electron transfer mediated by bacterial secreted flavins (Fig. 1.4) [23, 78]. The full electron pathway goes through cellular compartments by establishing an extensive network of redox events (Fig. 1.5). The oxidation of organic molecules takes place within the cytoplasm and electrons are transferred via a NADH dehydrogenase to IM soluble carriers (quinones). TorC and CymA are two periplasmic oriented tetraheme proteins anchored to the IM by a single α-helix that catalyze the regeneration of the quinones required for the continued cytoplasmic oxidation and transfer of the resulting electrons to a periplasmic redox partner. TorC and CymA are associated to different EET pathways depending on the terminal electron acceptor. In fact, it was found that TorC has as redox partner the TMAO reductase (TorA) and therefore is involved in the TMAO respiratory chain [79]. On the other hand, CymA is able to transfer electrons to partners that lead to the reduction of fumarate, nitrate, nitrite, mineral (hydr) oxides of Fe(III) and Mn(III/IV), DMSO, flavins or electrodes [80, 81].

Once the electrons are transferred to periplasmic cytochromes, these proteins can directly reduce the compounds that are able to cross the OM (*e.g.* sulfite, nitrate, nitrite and fumarate) or OM associated cytochromes that can then reduce extracellular compounds (such as DMSO, (hydr)oxides of Fe(III) and Mn(III/IV) and electrodes) [23, 82]. Recently, it was proposed that some OM proteins form porin-cytochrome complexes, providing a heme conduit for contiguous electron transfer across the OM [83, 84]. One of such examples is the MtrA-MtrB-MtrC complex identified in *S. oneidensis*. This complex is composed by three subunits; two of them are decaheme cytochromes *c* localized in the periplasm (MtrA) and in the bacterial cell surface (MtrC), whereas MtrB is an integral OM β-barrel protein that may serve as a sheath through which MtrA and MtrC might be inserted [85].

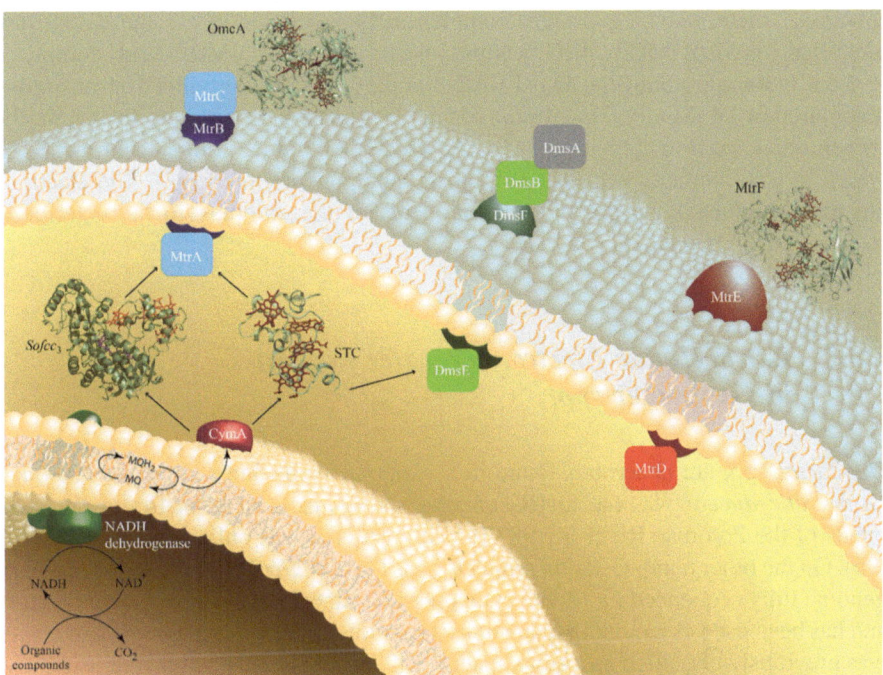

Fig. 1.5 Proposed model for extracellular electron transfer in *Shewanella oneidensis*. The cytoplasmic oxidation of organic molecules releases electrons to the menaquinone (MQ) pool via NADH dehydrogenase. From this point, a network of cytochromes *c* is responsible for the long-range electron transfer from the MQH_2 pool to extracellular acceptors. CymA is proposed to accept electrons from the MQH_2 pool and deliver them to multiheme periplasmic cytochromes, such as STC or *Sofcc₃*, which establish the interface between the cytoplasmic and the OM associated electron transfer components. Solid arrows connecting periplasmic components indicate the pairwise interactions identified by NMR experiments [86]. The cytochrome structures were drawn using the PyMOL molecular graphics system [156]: STC (PDB code 1M1Q [157]); *Sofcc₃* (PDB code 1D4D [19]); MtrF (PDB code 3PMQ [158]); OmcA (PDB code 4LMH [159]). Cartoons illustrate electron transfer components for which no structures are available: (i) OM complexes MtrA–MtrB–MtrC, MtrD–MtrE–MtrF and DmsEFAB, (ii) cytochrome CymA and (iii) NADH dehydrogenase. For recent reviews on *S. oneidensis* EET see Refs. [23, 86, 93, 118]

The current model for electron transfer pathway in *S. oneidensis* involving Mtr complexes proposes that electrons from organic source oxidation are transferred to CymA, which then transfer them to periplasmic electron transfer components, such as the small tetraheme cytochrome (STC) or flavocytochrome (*Sofcc₃*) that can ultimately interact with MtrA from the MtrA-MtrB-MtrC complex [86]. OmcA is a decaheme cytochrome co-expressed with MtrA-MtrB-MtrC and probably could receive electrons from the MtrC component. It was proposed that both OmcA and MtrC can act as reductases for external acceptors [87–92].

S. oneidensis genome contains two homologs of *mtrC* (*omcA* and *mtrF*), *mtrA* (*mtrD* and *dmsE*) and *mtrB* (*mtrE* and *dmsF*), some of which are also able to form functional metal-reducing protein complexes. For example, the same disposition identified for MtrA-MtrB-MtrC was proposed for the MtrD-MtrE-MtrF complex.

Less is known about the DmsEFAB complex, which is involved in anaerobic respiration of DMSO. In the current model, electrons are proposed to flow from the menaquinone pool to cytochrome CymA, which then reduces the cytochrome STC, which in turn interacts with the periplasmic component DmsE [93]. The electrons are then conducted via the other components of the complex to the exterior to reduce DMSO molecules [82, 94].

1.5.2   Extracellular Electron Transfer Pathways in Geobacter *spp.*

Geobacter spp. are anaerobes found in a variety of soils and sediments. Although *G. metallireducens* was the first *Geobacter* species isolated from freshwater sediments of the Potomac River in 1987 [95] it's genome was only sequenced in 2009 [96]. On the other hand, *G. sulfurreducens* was the first *Geobacter* species to have its genome fully sequenced by Methé et al. [56], it is amenable to genetic manipulation and has been used as a model organism for the study of *Geobacter*. *G. sulfurreducens* was previously classified as a strict anaerobe [97], but it was verified that it can also grow at low levels of atmospheric oxygen [98]. Additionally, the analysis of its genome revealed the presence of genes encoding typical enzymes for oxygen detoxification, such as catalase, cytochrome *c* oxidase or superoxide dismutase [98].

In natural environments where Fe(III) is an important respiratory terminal electron acceptor the microbial population is dominated by members of the *Geobacteraceae* family. However, their respiratory capacity is not confined to iron reduction since they can utilize a large diversity of electron donors and acceptors (Table 1.1) making them important agents in several biogeochemical cycles [58].

Until now, the EET pathways of *S. oneidensis* are better understood than the ones of *G. sulfurreducens*. As in the case of *S. onendensis*, the oxidation of organic compounds to carbon dioxide occurs in the cytoplasm and electrons are transferred to the menaquinone pool via the NADH dehydrogenase located at the IM. At this point, it was suggested that depending on the redox potential of the final electron acceptor different proteins are involved in the quinone regeneration [99, 100]. The cytochrome ImcH is an IM associated protein that contains seven heme binding motifs. Gene knock-out studies revealed that this cytochrome is required for respiration of extracellular electron acceptors with redox potentials higher than -0.1 V *versus* SHE (Fe(III) citrate, Fe(III)-EDTA, insoluble Mn(IV) oxides and electrodes poised at > 0.0 V *versus* SHE). However, ImcH is not essential for electron transfer to lower potential acceptors. Levar et al. [99] also showed that the growth rate of the cells at lower redox potentials is slower, which correlates with the fact that the cells generate less ATP when using lower-energy strategies. This observation also suggests that at least another quinone oxidoreductase is active in the bacteria to

Table 1.1 Respiratory versatility of representative *Geobacter* spp.

Name	Source	Electron donors oxidized with Fe(III)[a]	Fe forms reduced[b]	Other electron acceptors[a, c]
G. metallireducens	Aquatic sediments	Ac, Bz, Bze, BtOH, Buty, Bzo, BzOH, *p*-Cr, EtOH, *p*-HBz, *p*-HBzOH, IsoB, IsoV, Ph, Prop, PrOH, Pyr, Tol, Val	PCIO, Fe (III)-Cit	Mn(IV), Tc (VII), U(VI), AQDS, humics, nitrate
G. sulfurreducens	Contaminated ditch	Ac, H$_2$	PCIO, Fe (III)-Cit, Fe(III)-P	Tc(VII), Co (III), U(VI), AQDS, S°, Fum, Mal, O$_2$
G. bemidjiensis	Fe(III)-reducing subsurface sediment	Ac, Bzo, BtOH, Buty, EtOH, Fum, H$_2$, IsoB, Lac, Mal, Prop, Pyr, Succ, Val	Fe(III)-Cit, Fe (III)-NTA, Fe (III)-P, PCIO	AQDS, Fum, Mal, Mn(IV)
G. lovleyi	Freshwater sediment	Ac, Bze, Bzo, Buty, Cit, EtOH, For, Glu, Lac, MeOH, Prop, Succ, Tol, YE	Fe(III)-Cit, PCIO	PCE, TCE, nitrate, Fum, Mal, S°, U(VI), Mn(IV)
G. uraniireducens	Uranium-contaminated subsurface sediment	Ac, EtOH, Lac, Pyr	Fe(III)-NTA, Fe (III)-P, PCIO, smectite	AQDS, Fum, Mal, Mn(IV), U(VI)

Adapted from [58]

[a]Abbreviations for electron donors and acceptors: acetate (Ac), 9,10-anthraquinone-2,6-disulfonate (AQDS), benzaldehyde (Bz), benzene (Bze), benzoate (Bzo), benzylalcohol (BzOH), butanol (BtOH), butyrate (Buty), citrate (Cit), cobalt oxide (Co(III)), *p*-cresol (*p*-Cr), elemental sulfur (S°), ethanol (EtOH), formate (For), fumarate (Fum), glucose (Glu), *p*-hydroxybenzaldehyde (*p*-HBz), *p*-hydroxybenzylalcohol (*p*-HBzOH), hydrogen (H$_2$), isobutyrate (IsoB), isovalerate (IsoV), lactate (Lac), malate (Mal), methanol (MeOH), manganese oxide (Mn(IV)), oxygen (O$_2$) phenol (Ph), propanol (PrOH), propionate (Prop), pyruvate (Pyr), succinate (Succ), technetium oxide (Tc(VII)), tetrachloroethylene (PCE), trichloroethylene (TCE), toluene (Tol), uranyl acetate (U(VI)), valerate (Val), yeast extract (YE)

[b]Fe(III) forms: Poorly crystalline iron oxide (PCIO), ferric citrate (Fe(III)-cit), ferric nitrilotriacetic acid (Fe(III)-NTA), ferric pyrophosphate (Fe(III)-P)

[c]The bacteria has the ability to reduce the metal but it is not determined whether energy to support growth is conserved from the reduction of this metal

support growth at low redox potentials even when the *imcH* gene is deleted. Zacharoff et al. [100] found that the deletion of the gene *cbcL* that encodes for an IM HydC/FdnI diheme *b*-type cytochrome linked to a nonaheme periplasmic cytochrome *c* domain, severely affected the reduction of low potential electron acceptors, such as Fe(III)-oxides and electrodes poised at −0.1 V *versus* SHE. Identification of the quinone oxidoreductase CbcL supports an emerging model of multiple electron transfer pathways in *Geobacter*, where these bacteria rapidly alter their respiratory strategy to take advantage of the available extracellular electron acceptor. However, the electron transfer pathway from the menaquinone pool to the final electron acceptor is far to be understood.

Cytochrome MacA is a diheme cytochrome *c* (GSU0466) also associated to the IM. Sequence analysis of MacA showed homology to the diheme cytochrome *c* peroxidase CcpA (GSU2813), also from *G. sulfurreducens*. Cytochrome *c* peroxidases (Ccp) are enzymes that catalyze the reduction of hydrogen peroxide to water, and their presence indicates the occurrence of reactive oxygen species through metabolic activity or in microoxic environments. CcpA family proteins contain two *c*-type heme groups that differ in their axial ligands and, consequently, in their midpoint redox potential. One heme near the C-terminal is axially coordinated by His-Met and the other located near the N-terminal has a bis-histidinyl axial coordination and a significantly lower redox potential. Gene knock-out experiments revealed that the absence of the gene encoding for MacA affected the reduction of Fe(III) and U(VI) oxides [101, 102]. Additionally, MacA is more abundant during growth with Fe(III) oxides *versus* Fe(III) citrate [103].

PpcA belongs to a family composed by five low-molecular weight (∼ 10 kDa) periplasmic triheme cytochromes with approximately 70 residues each. The other four cytochromes of this family are designated PpcB, PpcC, PpcD, PpcE and share 77 % (PpcB), 62 % (PpcC), 57 % (PpcD) and 65 % (PpcE) amino acid sequence identity with PpcA. The heme groups in this family of cytochromes are axially coordinated by two histidines and have negative reduction potentials [104]. Genetic studies revealed that these cytochromes are involved in different electron transfer pathways. PpcA was detected in *G. sulfurreducens* cultures that grow in presence of Fe(III) citrate and Fe(III) oxide [103]. Deletion of the gene encoding for PpcA affected the reduction of Fe(III) and U(VI) [101, 105]. Genes encoding for PpcB and PpcC belong to the same locus and the double deletion of these genes affected U(VI) reduction. Both cytochromes were detected in Fe(III) citrate and Fe(III) oxide cultures [101, 103]. PpcD is more abundant during growth with Fe(III) oxides *versus* Fe(III) citrate [103]. The deletion of its gene affected the reduction of U(VI) [101]. PpcE was only detected in cultures with Fe(III) citrate and its deletion also affected U(VI) reduction [101]. Due to the cellular location of these five cytochromes it was proposed that they are the likely reservoir of electrons destined for the cell outer surface and bridge the electron transfer between the cytoplasm and the exterior in *G. sulfurreducens* [105, 106].

Also located in the periplasm of *G. sulfurreducens* the dodecaheme cytochrome GSU1996 (42.3 kDa) is composed by four similar triheme domains, designated A, B, C and D (see Sect. 1.6.1). Proteomic studies showed that this protein is more abundant during growth with Fe(III) citrate *versus* Fe(III) oxide [103] and the twelve-heme groups might allow the cytochrome to act as a periplasmic electron capacitor [25].

In order to reduce extracellular compounds, the electrons are transferred from the periplasm to the OM proteins. As suggested for *S. oneidensis*, the electrons cross the OM through porin-cytochrome trans-outer membrane complexes. The complexes OmaB-OmbB-OmcB and OmaC-OmbC-OmcC in *G. sulfurreducens* may have similar functions to the ones described for the MtrA-MtrB-MtrC complex in *S. oneidensis* [107]. Theses complexes consist of a porin-like outer membrane protein (OmbB or OmbC), a periplasmic octaheme cytochrome *c* (OmaB or OmaC) and an OM dodecaheme cytochrome *c* (OmcB or OmcC). Studies revealed that these complexes are directly involved in the reduction of Fe(III)-citrate and ferrihydrite by *G. sulfurreducens* [108].

OmcF (GSU2432) is the smallest monoheme cytochrome *c* (11 kDa) of *G. sulfurreducens* predicted to be localized at the OM [56]. The heme group is axially coordinated by His-Met [109], both in the reduced and oxidized states. Kim et al. [110] showed that the deletion of the *omcF* gene impaired Fe(III) citrate reduction and affected the expression of the outer membrane MHC OmcB, OmcC and OmcS. Specifically, deletion of *omcF* resulted in a loss of expression of *omcB* and *omcC*, and overexpression of *omcS*, during growth on Fe(III) citrate [111]. More recently, Aklujkar et al. [59] demonstrated that a OmcF-deficient strain was unable to grow in presence of Fe(III) oxide but no effect was observed in the presence of Mn(IV) oxides. This strain also showed an important decrease in current production [111].

OmcS and OmcZ are two major extracellular *c*-type cytochromes in *G. sulfurreducens* [112, 113]. It was demonstrated that OmcS is associated with conductive pili [113] and OmcZ possibly functions as an intermediary in electron flow to pili, which could serve as the ultimate conduit for long-range electron transfer [114]. OmcS cytochrome (47 kDa) contains six heme groups. The deletion of the gene encoding this cytochrome affected the bacterial growth in the presence of Fe (III) and Mn(IV) oxides [115]. It was also observed that OmcS is more abundant during growth with Fe(III) oxides *versus* Fe(III) citrate [114]. The opposite effect was observed for OmcZ, a 30 kDa octaheme cytochrome [103, 112]. The expression of *omcZ* also appears to downregulate the expression levels of *omcS*. Furthermore, *omcZ* inactivation inhibited current production and biofilm formation [114]. The structures of OmcS and OmcZ are not yet available and their heme content was determined by pyridine hemochrome assays. Details on the heme axial coordination were obtained from the analysis of their spectroscopic features in solution, which suggested that they are axially coordinated by two histidine residues [116, 117].

1.6 Characterization of Multiheme Cytochromes

The key role of MHC with considerable molecular weight and several redox centers in EET urge for the development of methodologies that facilitate their characterization. Some MHC are membrane associated and their characterization presents additional challenges because the solution studies by nuclear magnetic spectroscopy (NMR) requires the incorporation of detergent micelles, which increase considerably the overall molecular weight and introduce intense extra sharp peaks in the spectra [118]. Therefore, the scaling-up of the existent methodologies or the implementation of new ones that underlying the structural and functional characterization of MHC would contribute to the understanding of the overall EET processes and to establish foundations that can be explored to optimize EET-based biotechnological applications.

The structural and functional characterization of MHC is, therefore, crucial to understand the EET mechanisms and to engineer improved forms of these electron transfer components that ultimately will contribute to increase the bacterial respiratory rates. In this Section we focus on the methodologies currently applied to characterize MHC using mostly the triheme cytochromes from *G. sulfurreducens* as models.

1.6.1 Structural Characterization

Three-dimensional structures of biological macromolecules can be determined by X-ray crystallography and NMR at near atomic resolution. The first approach has no size limitations but it requires the formation of diffracting protein crystals. On the other hand, NMR is suitable for the characterization of macromolecules in solution with molecular weights generally below 50 kDa, although several efforts have been carried out to increase the protein molecular size limits. The structures obtained by NMR represent an average over semi-randomly oriented molecules in solution, whereas diffraction data represent an average over molecules arranged in a periodic crystal lattice. Despite their similarities and differences, X-ray crystallography and NMR are well established complementary high resolution methods to analyze protein structure-function relationships.

The main information used to determine the structure of a protein in solution relies on proton-proton distance restraints, obtained from nuclear Overhauser effect (NOE) signal intensities correlating nuclei in close spatial proximity (up to ~ 5 Å). These restraints are obtained from the analysis of 2D ^{1}H-Nuclear Overhauser Spectroscopy (NOESY) NMR experiments, which require an unambiguous assignment of the protein signals. However, in the case of MHC this assignment is not straightforward, particularly in the paramagnetic state, due to the magnetic properties of the heme iron that further spread and broad the NMR signals [119, 120]. In fact, each heme includes several proton-containing groups (four methyl groups, four meso protons, two thioether protons, two thioether methyls and two propionates; see Fig. 1.1) that contribute to the NMR signal ensemble. Moreover, due to the proximity of these groups and those of the polypeptide chain, a large number of NOE connectivities involving both types of signals are also present in the NMR spectra that must be unambiguously assigned to provide adequate restraints for solution structure determination. Consequently, the MHC

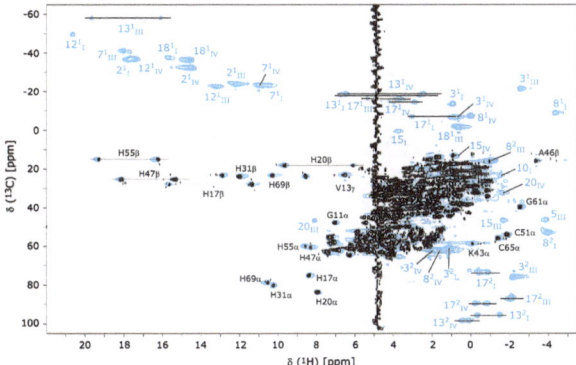

Fig. 1.6 2D^{1}H,^{13}C HSQC NMR spectra of PpcA samples (298 K and pH 5.5). The unlabeled sample and sample labeled exclusively on the polypeptide chain are represented by *blue* and *black contours*, respectively. The resonances shifted from the main ensemble of signals are labeled. *Blue* and *black labels* indicate the heme substituents and the polypeptide resonances, respectively. The signals of the protons connected to the same carbon atom (CH$_2$ groups) are linked by a line. This Figure was reproduced with permission from Elsevier, reference [120]

superfamily is extremely under-represented in structural databases, which constitutes a severe bottleneck in the elucidation of their structural–functional relationships. Advances in protein expression protocols have contributed to increase the expression yields for mature MHC [26, 72, 73, 121, 122]. These made the isotopic labeling of MHC more cost-effective [123] and contributed significantly to overcome the traditional difficulties associated with the determination of solution structures using natural abundance samples. A methodology that enables the isotopic labeling of MHC exclusively in their redox centers was reported [124] and is expected to be a valuable tool in the assignment of the NMR heme signals in very large MHC for which the spectral overlap is severe. A strategy that simplifies the assignment of these signals was developed by Morgado et al. [120]. This combines the analysis of ^{1}H,^{13}C heteronuclear single-quantum coherence (HSQC) NMR experiments obtained for an unlabeled sample and for a sample labeled exclusively in its polypeptide chain. The simple comparison of these spectra allows a straightforward discrimination between the heme and the polypeptide chain signals and is illustrated for PpcA in Fig. 1.6. Taking advantage of the methodologies described above, the NMR fingerprints, including heme proton, protein backbone and side chain NH signals, were identified for several cytochromes of the PpcA family, in both oxidized and reduced forms [125–127], establishing important basis to determine their structure and dynamic properties in solution [128].

Out of MHC found in *Geobacter*, the best characterized to date are the five periplasmic triheme cytochromes from the PpcA family. The structures of these cytochromes were determined by X-ray crystallography in the oxidized state and revealed a high structural homology [129–131]. The solution structure of PpcA in the reduced form was also determined by NMR [128]. The tertiary structure of all

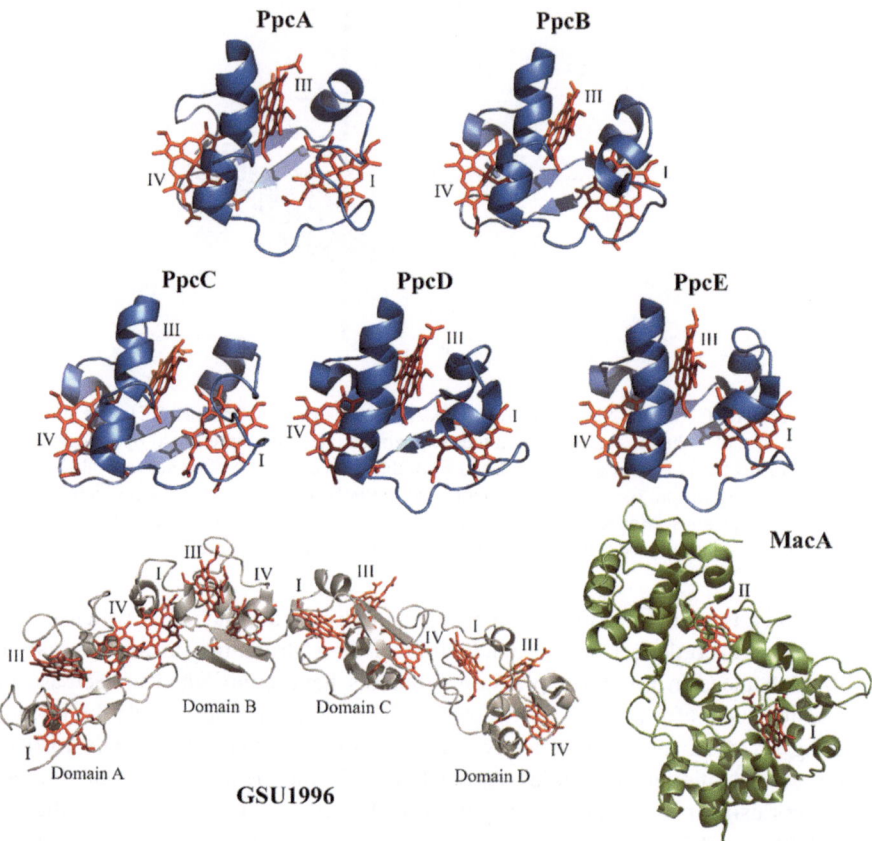

Fig. 1.7 Structures of *G. sulfurreducens* multiheme cytochromes. The solution structure of PpcA (lowest energy; PDB code 2LDO [128] and the crystal structures of PpcB (PDB code 3BXU [130]), PpcC (PDB code 3H33 [131]), PpcD (PDB code 3H4N [131]), and PpcE (PDB code 3H34 [131]) are represented in *blue*. Crystal structures of GSU1996 (PDB code 3OV0 [25]) and MacA (PDB code 4AAL [133]) are represented in *gray* and *green*, respectively. Roman numerals indicate the hemes (in *red*) in their order of attachment to the Cys-X-X-Cys-His motif in the polypeptide chain. The structures were drawn using the PyMOL molecular graphics system [156]

proteins is similar, although local variations were observed. As depicted in Fig. 1.7, an antiparallel β-sheet is conserved at the N-terminal of all structures, and is followed by distinct helical regions in the different proteins. The heme core structures are similar, with hemes I and IV roughly parallel to each other and with both of them nearly perpendicular to heme III [128–131]. Within the five structures, in general heme I shows the highest solvent exposure, while heme III is the less exposed. On the other hand, heme IV shows the largest positive electrostatic surface due to the considerable number of neighboring lysine residues. This positively charged surface around heme IV is the most conserved region, while the lowest similarity region is located near heme I [131]. The spatial arrangement of the hemes is superimposable with those of the structurally homologous tetraheme cytochromes c_3, with the sole difference being the absence of heme II and the corresponding polypeptide segment. For this reason, the three heme groups in triheme cytochromes are numbered I, III and IV [132].

Structures for other *G. sulfurreducens* cytochromes were also determined by X-ray crystallography, namely those of cytochrome GSU1996 in the oxidized state [25] and MacA in the reduced, oxidized and intermediate states [133] (Fig. 1.7).

GSU1996 is arranged in a "crescent-shaped" architecture forming a "nanowire" of hemes that may function as an electron-storage sink or capacitor to enhance the periplasmic bacterial electron-storage capacity. It is organized into four domains (A-D), each containing three hemes with structural homology to triheme cytochromes, differing only in heme IV, which presents His-Met axial coordination. Despite the presence of four hemes with His-Met axial coordination the heme reduction potential of GSU1996 is negative [134].

The crystal structure of MacA was determined in the reduced, oxidized and intermediate states [133]. MacA forms two globular domains, each holding a *c*-type heme group. The hemes were named as high potential (HP) and low potential (LP) and there is no electron transfer between them [133]. The LP heme is located at the N-terminal domain whereas the C-terminal domain harbors the HP one. The HP heme is reduced by the physiological electron donor whereas the LP heme is the site for hydrogen peroxide reduction. Altogether, the studies indicated that the functional mechanism of MacA resembles that of proteins of the CcpA family [133]. In fact, the reduction of the HP heme triggers a concerted conformational rearrangement of three loop regions that opens up a free coordination site at the distal axial position of the LP heme group, allowing the access of the substrate to the active site. In addition, it was also shown that MacA mediates the electron transfer to the periplasmic triheme cytochrome PpcA (GSU0612) by electrochemical experiments [133]. According to these authors, MacA might have a bi-functional behavior that includes peroxidase activity or act as redox partner of PpcA.

1.6.2 Thermodynamic Characterization

As for the solution structure determination, the presence of several heme groups constitutes the major bottleneck in the detailed thermodynamic characterization of MHC. Indeed, while for monoheme cytochromes only the reduced and oxidized states may coexist in solution, in MHC several one-electron reversible transfer steps convert the fully reduced state into the fully oxidized yielding additional intermediate oxidation stages. Therefore, the global redox potentials measured by voltammetry or potentiometric redox titrations are macroscopic in nature and only describe the overall redox behaviour of the MHC [109, 116, 117, 135]. Although these parameters contain information on the working functional ranges of these cytochromes, in most cases they are insufficient to provide mechanistic information on the electron transfer pathways. This can only be achieved when the fractional contribution of each possible microstate during the oxidation of the protein is obtained (see below).

Another important aspect of the functional characterization of multiheme cytochromes is the kinetics of intermolecular electron transfer. This is measurable by combining the microscopic thermodynamic characterization of the redox centers with stopped-flow measurements. As for the redox potentials, the analysis of kinetic data requires the distinction between the observable macroscopic rate constants and the structurally relevant microscopic properties. A model that analyzes the electron transfer kinetics between multicenter redox proteins, their electron donors and acceptors and, therefore allows to obtain microscopic information on the protein kinetic properties, was originally described by Catarino and Turner [136]. This model was recently revisited by Paquete and Louro [137] and its description is beyond the scope of this book.

Thermodynamic Parameters
The detailed thermodynamic characterization of MHC requires the determination of the heme reduction potentials and other parameters (see below). In the case of monoheme cytochromes the reduction potential of the heme can be obtained directly from the Nernst equation. In this case, the e_{app} value (i.e., the point at which the oxidized and reduced fractions of the protein are equal) corresponds to the reduction potential of the heme. However, this is not the case for MHC. In fact, for a MHC containing N heme groups, the same number of one-electron reversible transfer steps converts the fully reduced protein into the fully oxidized one. Therefore, the Nernst equation does not describe accurately such system and even a model that considers sequential midpoint reduction potentials for each center will only provide values that correspond to macroscopic redox potentials, which cannot be formally assigned to any specific heme. Indeed, as a consequence of the close disposition of the heme groups in MHC the redox potential of one heme is modulated by the oxidation state of a neighboring one (redox interactions). Often, the redox potential of the hemes are also modulated by the pH (redox-Bohr effect). The magnitude of this effect is determined by another set of parameters designated redox-Bohr interactions, which quantify the effect of the protonation state of the redox-Bohr center on the heme reduction potentials. Therefore, to achieve a complete thermodynamic characterization of a MHC, besides the heme redox potentials, it is also necessary to determine the redox interactions and the properties of the redox-Bohr center(s).

The experimental determination of such parameters requires the monitorization of the oxidation profile of each heme group. Optical spectroscopies typically fail in such distinction as the hemes with the same axial ligands have nearly identical spectral signatures. On the other hand, in most cases, NMR can be used for this purpose (see below).

   A model that allows the determination of MHC thermodynamic parameters was proposed by Turner et al. [138] and was used for the first time to characterize the tetraheme cytochrome c_3 from *Desulfovibrio vulgaris* (Hildenborough). In this book the model is revisited by considering a triheme cytochrome with one redox-Bohr center. In this case, four oxidation stages linked by successive one-electron reductions can be defined (Fig. 1.8a). Each oxidation stage, numbered S_0–S_3, comprises all the microstates with the same number of oxidized hemes. The fractional contribution of the 16 microstates can be defined by 10 thermodynamic parameters for any pH and redox potential value: three reduction potentials, three redox interactions between each pair of hemes, three redox-Bohr interactions between the hemes and the redox-Bohr center and the pK_a of this center (Fig. 1.8b). The energy of each microstate relatively to the fully reduced and protonated protein is then given by a simple sum of the appropriate energy of the four independent centers and the six possible two-center interactions, plus a *SFE* term, which accounts for the effect of the solution potential, E, in the oxidation stage S, and another for the proton chemical potential 2.3RTpH, added for the deprotonated forms.

   Therefore, the energy (G) of a microstate relative to the reference microstate (fully reduced and protonated) is given by Eq. 1.1:

$$G_{mH} = \Sigma g_i + \Sigma g_{i-j} + SFE \qquad (1.1)$$

where mH designate a particular protonated microstate m, i the heme group, g_i the energy of oxidation of heme i, g_{i-j} the interaction energy between each pair of hemes, S the oxidation stage (that correspond to the number of oxidized hemes) and E the redox potential of the solution. In the case of microstates with the redox-Bohr center deprotonated, the energy is given by Eq. 1.2:

$$G_m = G_{mH} + g_H + \Sigma g_{iH} + 2.3(RT/F)pH \qquad (1.2)$$

where g_H corresponds to the deprotonation energy of the fully reduced protein and g_{iH} to the energy of interaction between the hemes and the redox-Bohr center.

   The energy values can be converted to reduction potentials using Eq. 1.3:

$$\Delta G \;=\; \text{-}nF\Delta E \qquad (1.3)$$

and the fractional contribution of each microstate (P_m) can be determine by the Boltzmann equation (Eq. 1.4):

$$P_m = \exp((\text{ -}F/RT)G_m) \qquad (1.4)$$

Experimental Data

To determine the thermodynamic parameters of a MHC, it is necessary to monitor the oxidation profile of the hemes at several pH values by 2D-exchange NMR spectroscopy (EXSY) experiments. This can only be achieved when the electron

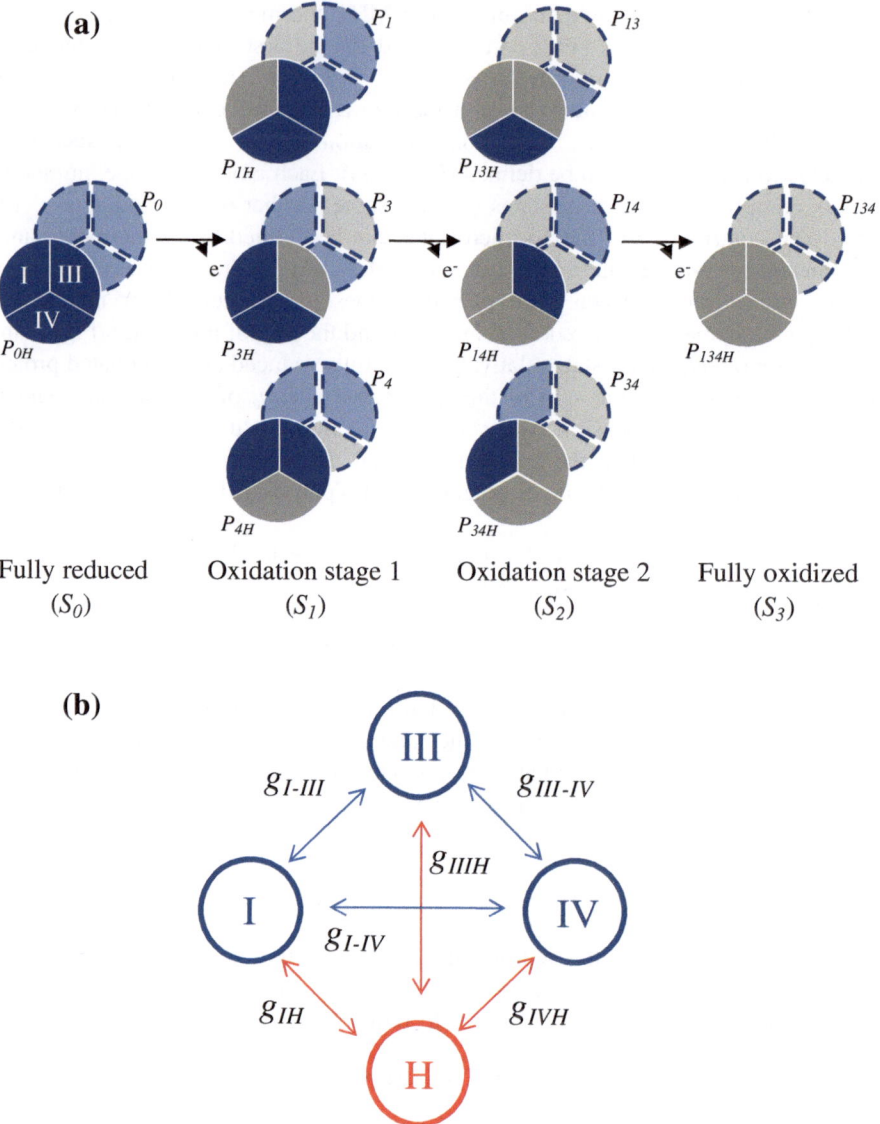

Fig. 1.8 **a** Electronic distribution scheme for a triheme cytochrome with a proton-linked equilibrium showing the 16 possible microstates. The *dark solid* and *light dashed circles* correspond to the protonated and deprotonated microstates, respectively. Each *circle* is divided into three parts that represent the heme groups, which can be either reduced (*blue*) or oxidized (*gray*). The microstates are grouped, according to the number of oxidized hemes, in four oxidation stages connected by three one-electron redox steps. P_{0H} and P_0 represent the reduced protonated and deprotonated microstates, respectively. P_{ijkH} and P_{ijk}, indicate respectively the protonated and deprotonated microstates, where i, j, and k represent the heme(s) that are oxidized in that particular microstate. For multiheme cytochromes containing higher number of hemes (N) and redox-Bohr centers (NB), the distribution can be easily scaled-up. The total number of microstates and oxidation stages would be given by $2^N(NB + 1)$ and $N + 1$, respectively. **b** Schematic representation of the interaction network of a MHC with three hemes (numbered I, III, IV) and one redox-Bohr center (H). The interacting energies between the centers are represented as g_{ij} and g_{iH}, where i and j represent a pair of hemes

exchange is fast between the different microstates within the same oxidation stage (intramolecular) and slow between the different oxidation stages (intermolecular) on the NMR time scale [130]. Optimization of the experimental conditions to favour slow intermolecular electron exchange regime include (i) the increase of the solution ionic strength, (ii) the lowering of the temperature, (iii) the decrease of the protein concentration and (iv) the use of higher magnetic fields [104]. When fast intra- and slow intermolecular electron exchange are meet on the NMR time scale, the heme oxidation fractions can be determined from the chemical shifts of their heme substituents in the different oxidation stages. The heme methyl resonances are the easiest identifiable NMR signals amongst the heme substituents, making them ideal candidates to monitor the stepwise oxidation of the hemes. In fact, as depicted in Fig. 1.9 and Table 1.2, the chemical shift of the heme signals are considerably different in the reduced and oxidized states, shifting from crowded regions in the fully reduced to relatively empty regions in the fully oxidized spectra (Fig. 1.9). Therefore, the signals can be followed during a redox titration monitored by 2D ^{1}H-EXSY NMR spectra as illustrated in Fig. 1.10 for the heme methyls of cytochrome PpcA.

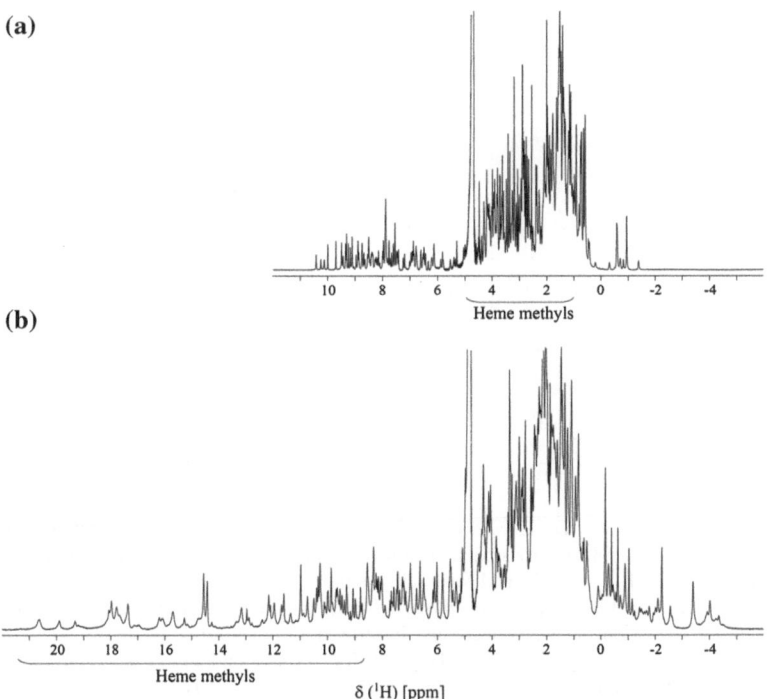

Fig. 1.9 1D ^{1}H-NMR spectra of the reduced (**a**) and oxidized (**b**) triheme cytochrome PpcA (298 K and pH 5.5). The typical regions of the heme methyl substituents are indicated. This Figure was reproduced with permission from Elsevier, reference [120]

Table 1.2 Chemical shifts of the heme methyl protons of PpcA in the fully reduced and oxidized states (pH 7.1 and 298 K) [126, 150]

Heme methyl	Chemical shifts (ppm)		
	Heme I	Heme III	Heme IV
2^1CH_3	3.54 [17.79]	4.33 [12.15]	3.59 [14.93]
7^1CH_3	3.56 [10.60]	4.13 [18.09]	3.01 [10.61]
12^1CH_3	2.54 [21.09]	3.49 [13.18]	3.93 [19.12]
18^1CH_3	3.33 [15.78]	3.84 [0.79]	3.33 [14.66]

The chemical shifts of PpcA in the fully oxidized state are indicated in parenthesis

The paramagnetic shifts of the heme methyls are proportional to the oxidation fraction of a particular heme and, thus contain information that can be used to infer about the redox properties of each heme group [139, 140]. However, the NMR data per se are insufficient to determine the absolute thermodynamic parameters and need to be complemented with data from potentiometric redox titrations monitored

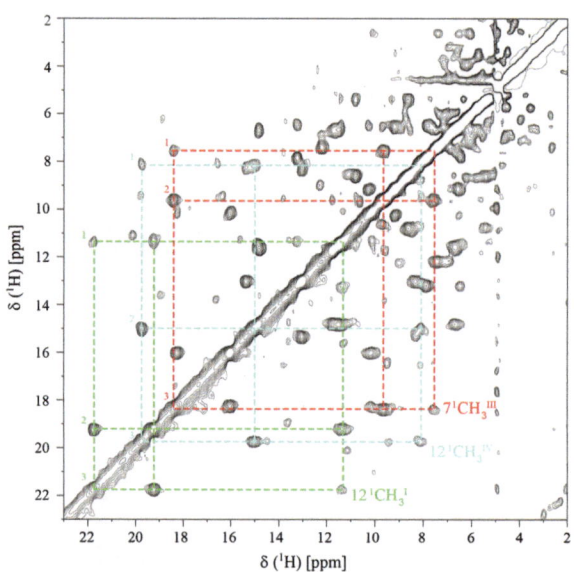

Fig. 1.10 Expansions of 2D ^{1}H-EXSY NMR spectra obtained for PpcA at different levels of oxidation (288 K and pH 8). Cross-peaks resulting from intermolecular electron transfer between the oxidation stages 1–3 are indicated for the heme methyls $12^1CH_3^I$ (*green dashed lines*), $7^1CH_3^{III}$ (*red dashed lines*) and $12^1CH_3^{IV}$ (*blue dashed lines*). Roman and Arabic numbers indicate the hemes and the oxidation stages, respectively. In order not to overcrowd the figure, the cross-peaks to oxidation stage 0 are not shown. The chemical shifts correspondent to the oxidation stage 0 for each heme methyl are: 2.55; 4.14 and 3.95 ppm for $12^1CH_3^I$; $7^1CH_3^{III}$ and $12^1CH_3^{IV}$, respectively. This research was originally published in Bioscience Reports. Morgado L, Dantas JM, Simões T, Londer YY, Pokkuluri PR, Salgueiro CA., Role of Met(58) in the regulation of electron/proton transfer in trihaem cytochrome PpcA from *Geobacter sulfurreducens*. Bioscience Reports. 2012; 33:11-22 © copyright holder

by visible spectroscopy, obtained for at least two different pH values [138]. Owing to the negative reduction potentials of the hemes, in most of the cases, these redox titrations are performed inside an anaerobic glove chamber. The titrations are carried out in the presence of redox mediators that assist in transferring electrons between the electrode and the protein redox centers. In order to confirm the reversibility of the process the redox titrations should be performed in the oxidative and reductive directions. For each solution potential, the correspondent UV-visible spectrum is recorded and the reduced fraction of the proteins determined upon normalization of the integrated area of the α-band above the line connecting the flanking isosbestic points to subtract the optical contribution of the redox mediators (see inset in Fig. 1.3). The α-band is characteristic of each cytochrome c and can vary between 549 and 556 nm [43].

The full set of NMR and visible data used to determine the thermodynamic parameters of PpcA (Table 1.3) is indicated in Fig. 1.11. The quality of the fittings obtained for the pH-dependence of the paramagnetic chemical shifts and for the

Table 1.3 Thermodynamic parameters for triheme cytochrome PpcA from *G. sulfurreducens*; tetraheme cytochrome c_3 from *D. vulgaris* (Hildenborough) (*Dv*Hc$_3$) and flavocytochrome c_3 from *S. frigidimarina* (*Sffcc$_3$*)

Energies (meV)					
	Heme I	Heme II	Heme III	Heme IV	Redox-Bohr center
PpcA [104]					
Heme I	**−154 (5)**		27 (2)	16 (3)	−32 (4)
Heme II		N/A	N/A	N/A	N/A
Heme III			**−138 (5)**	41 (3)	−31 (4)
Heme IV				**−125 (5)**	−58 (4)
Redox-Bohr center					**495 (8)**
***Dv*Hc$_3$** [151]					
Heme I	**−252 (2)**	−39 (1)	19 (1)	3 (3)	−74 (3)
Heme II		**−284 (2)**	1 (1)	10 (2)	−36 (2)
Heme III			**−343 (1)**	34 (2)	−23 (2)
Heme IV				**−293 (2)**	−14 (2)
Redox-Bohr center					**454 (3)**
Sffcc$_3$ [147]					
Heme I	**−228 (5)**	23 (3)	3 (3)	−2 (6)	−10 (4)
Heme II		**−270 (5)**	65 (2)	0 (3)	−10 (4)
Heme III			**−223 (6)**	12 (2)	−11 (4)
Heme IV				**−148 (8)**	−10 (4)
Redox-Bohr center					**504 (36)**

In each case, the fully reduced and protonated protein was taken as reference. Diagonal values (in bold) correspond to oxidation energies of the hemes and deprotonating energy of the redox-Bohr center. Off-diagonal values are the redox and redox-Bohr interaction energies. All energies are reported in meV, with standard errors given in parenthesis. N/A—not applicable

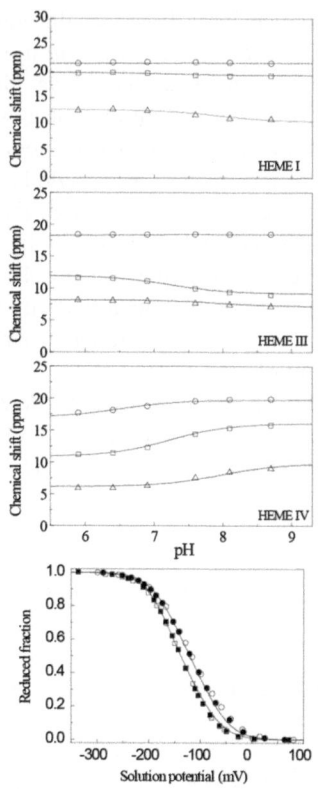

Fig. 1.11 Fitting of the thermodynamic model to the experimental data for PpcA. The solid lines are the result of the simultaneous fitting of the NMR and visible data. The three upper panels show the pH dependence of heme methyl chemical shifts at oxidation stages 1 (*open triangle*), 2 (*open square*), and 3 (*open circle*). The chemical shift of the heme methyls in the fully reduced stage (stage 0) are not plotted since they are unaffected by the pH. The bottom panel corresponds to the reduced fractions of the various cytochromes, determined by visible spectroscopy at pH 7 (*circles*) and pH 8 (*squares*). The *open* and *filled symbols* represent the data points in the reductive and oxidative titrations, respectively. This Figure was reproduced with permission from Elsevier, reference [104]

visible redox titrations clearly shows that the experimental data is well described by the model. Once the thermodynamic parameters are determined it is possible to evaluate the contribution of each microstate during the oxidation of the protein and infer about the mechanistic features of the MHC under study.

Functional Mechanisms and Implications

The detailed thermodynamic characterizations of the triheme cytochrome from *G. sulfurreducens* (PpcA), the tetraheme cytochrome c_3 from *D. vulgaris* (Hildenborough) (*DvHc₃*) and the flavocytochrome c_3 from *S. frigidimarina* (*Sffcc₃*) are presented in this Section. These examples illustrate the variability of the thermodynamic parameters within MHC and reveal their versatile functional

Table 1.4 pK_a values of the redox-Bohr center for PpcA [104], $DvHc_3$ [151] and $Sffcc_3$ [147] in the different oxidation stages

Oxidation stage	pK_a		
	PpcA	$DvHc_3$	$Sffcc_3$
0	8.6	7.7	8.5
1	8.0	7.2	8.4
2	7.2	6.5	8.2
3	6.5	5.6	8.0
4	N/A	5.2	7.8
ΔpK_a	2.1	2.5	0.7

The pK_a values correspondent to the reduced and oxidized states are given by $g_B F/(2.3RT)$ and $(g_B + \sum_{i=1}^{3} g_{iB})F/(2.3RT)$, respectively. The values $g_B F$ and g_{iB} are listed in Table 1.3. N/A—not applicable

mechanisms. The thermodynamic parameters and the pK_a values of the redox-Bohr centers of the three cytochromes are indicated in Tables 1.3 and 1.4, respectively. The analysis of Table 1.3 shows that for each cytochrome, the reduction potentials of the hemes are negative, different from each other and cover different potential ranges.

Except for the tetraheme cytochrome c_3, the redox interactions are in general positive, indicating that the oxidation of a particular heme renders the oxidation of its neighbors more difficult. Also, the strongest redox interactions are observed between the closest pair of heme groups, which is expected for charged centers with a Coulombic behavior.

Regarding the redox-Bohr interactions, they are in general negative, i.e., the oxidation of the hemes facilitates the deprotonation of the redox-Bohr center and *vice versa*. In structural terms, the more negative redox-Bohr interaction, the closer proximity of the redox-Bohr center to that particular heme group. The magnitude of the redox-Bohr interactions are quite variable amongst the various MHC and can even be not significant, as in the case of flavocytochrome c_3. The data obtained for PpcA and $DvHc_3$ indicates that the redox-Bohr center is closer to hemes IV and heme I, respectively. The magnitude of the redox-Bohr effect is also reflected in the difference between the pK_a values obtained for the reduced and oxidized proteins. The larger redox-Bohr effects ($\Delta pK_a = pK_a^{red} - pK_a^{ox}$) are observed for $DvHc_3$ and PpcA (2.5 and 2.1 pH units, respectively). The ΔpK_a value is considerably smaller for $Sffcc_3$ (0.7 pH units) (Table 1.4).

Order of oxidation of the heme groups

The relative order of oxidation of the heme groups in the fully protonated and reduced state can be obtained from the values of the microscopic reduction potentials listed in Table 1.3 and corresponds to I–III–IV; III–IV–II–I and II–I–III–IV for PpcA, $DvHc_3$ and $Sffcc_3$, respectively. However, as mentioned above, the reduction potential of each heme is affected by the oxidation state of neighboring ones (redox interactions) and by the pH (redox-Bohr interactions). This modulation

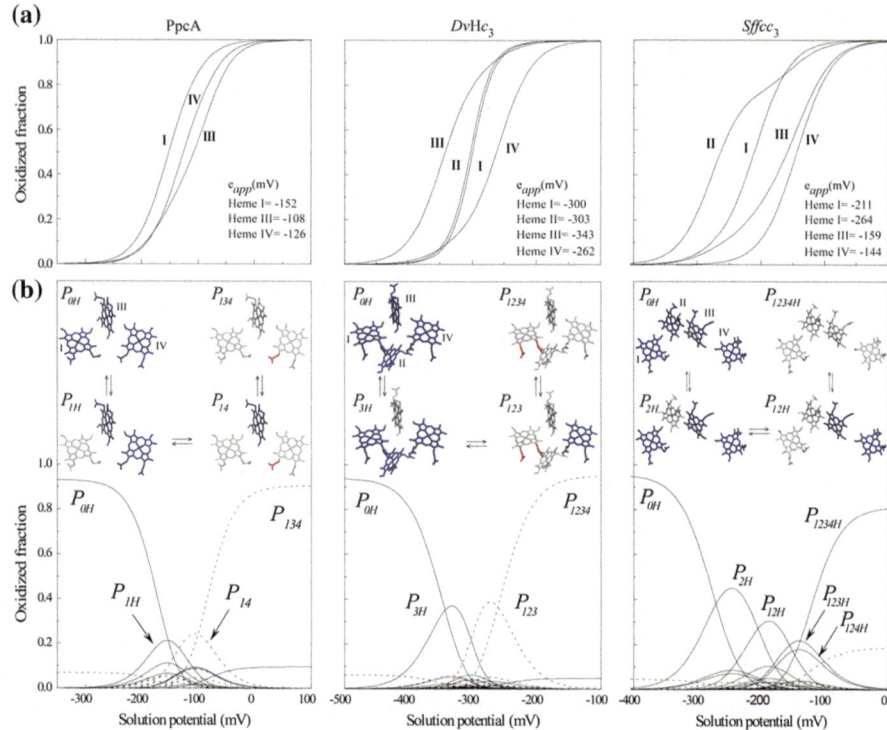

Fig. 1.12 a Oxidized fractions of the individual hemes for PpcA, $DvHc_3$ and $Sffcc_3$ at physiological pH. The curves were calculated as a function of the solution reduction potential using the parameters listed in Table 1.3. The midpoint reduction potentials (e_{app}) of the hemes are indicated. **b** Molar fractions of the individual microstates at physiological pH. The curves were calculated as a function of the solution reduction potential using the parameters listed in Table 1.3. *Solid* and *dashed lines* indicate the protonated and deprotonated microstates, respectively. For clarity only the relevant microstates are labeled. The dominant microstates are also illustrated by the heme core of the proteins. Reduced and oxidized hemes are colored *blue* and *gray*, respectively. The deprotonated redox-Bohr center is colored *red*. Roman numerals indicate the hemes in their order of attachment to the Cys-X-X-Cys-His motif in the polypeptide chain. The structures (PpcA (PDB code 2LDO [128]); $DvHc_3$ (PDB code 2CTH [154]) and $Sffcc_3$ (PDB code 1QJD [20]) were drawn using the PyMOL molecular graphics system [156]

is reflected in the individual oxidation profiles of the hemes shown in Fig. 1.12a calculated for the physiological pH region of each bacterium. These oxidation curves are substantially different from a Nernst curve, and the several crossovers between them clearly indicate that the electron affinity of each heme is tuned by redox interactions with neighboring ones. As a consequence, the midpoint reduction potentials e_{app} differ from those observed in the fully reduced protein (cf. Table 1.3 and Fig. 1.12a). Consequently, the actual order of the oxidation of the hemes at physiological pH is I–IV–III; III–(II,I)–IV and II–I–III–IV for PpcA, $DvHc_3$ and

Sffcc₃, respectively. It is worth nothing that in the case of *Sffcc₃* the order of oxidation of the hemes is the same at physiological pH, as the redox-Bohr inter-actions are not significant (Table 1.3).

Relevant microstates in solution and functional implications

The molar fractions of the microstates at physiological pH can be also obtained from the thermodynamic parameters listed in Table 1.3 (Fig. 1.12b). In the case of PpcA, the oxidation stages 0 and 1 are dominated by the protonated forms P_{0H} and P_{1H}, respectively, while the redox-Bohr center is kept protonated. Stage 2 is dominated by the oxidation of heme IV and deprotonation of the redox-Bohr center (P_{14}), which remains deprotonated in stage 3 (P_{134}). Therefore, a route for the electrons is defined within PpcA: $P_{0H} \rightarrow P_{1H} \rightarrow P_{14} \rightarrow P_{134}$ suggesting that the protein has the thermodynamic properties necessary to facilitate e^-/H^+ energy transduction. The notorious redox-Bohr effect observed for PpcA at physiological pH (pH 7.5) could be important for the events that contribute to establish the proton electrochemical potential gradient across the periplasmic membrane. In fact, the oxidation progresses from a specific protonated redox-microstate (P_{1H}) to a par-ticular deprotonated redox-microstate (P_{14}), showing how dominant microstates can confer the directionality of events (Fig. 1.12b). Therefore, PpcA can uptake a strongly reducing electron (-167 mV) and a weakly acidic proton (pK_a 8) from a donor associated with the cytoplasmic membrane. When it meets its physiological downstream redox partner, PpcA donates de-energized electrons (-109 mV) and lowers the pK_a of the proton (pK_a 7.2), which can be more efficiently released into the periplasm.

Analysis of the molar fraction of the microstates in *DvHc₃* clearly points to the existence of a proton-assisted two-electron step, as the contribution from micro-states belonging to the oxidation stage 2 is significantly decreased compared to those of oxidation stages 1 and 3. Therefore, in this cytochrome the preferential route for the electrons is $P_{0H} \rightarrow P_{3H} \rightarrow P_{123} \rightarrow P_{1234}$. In this case, the redox-Bohr center deprotonates with the concerted transfer of two electrons from P_{3H} to P_{123}. The concerted transfer of $2e^-$ is also illustrated by the steep slope and by the close proximity of the oxidation curves of hemes I and II, which results from their strong negative redox interaction (cf. Table 1.3 and Fig. 1.12a). The negative value of this interaction indicates that the transfer of one electron facilitates the transfer of a second one, which is unexpected in electrostatic terms. Therefore, the anti-Coulombic value of the redox interaction between hemes I and II points to the existence of an important redox-linked structural conformational change between oxidation stages 1 and 3. Combined, these results clearly show that *DvHc₃* can perform a $2e^-/2H^+$ concerted step at physiological pH. Thus, this cytochrome is particularly well designed to work as a coupling protein to its physiologic partner, the enzyme hydrogenase. In fact, in the presence of sulfate, *Desulfovibrio* spp. can use H_2 as the sole energy source, which can be oxidized in the periplasm ($H_2 \rightarrow 2H^+ + 2e^-$) [141–143]. The finding that two protons are involved in the redox-Bohr effect of *DvHc₃* [144] and that the protein is also designed to perform a $2e^-/2H^+$

concerted step [138] supports its central role in the energy transduction mechanisms of sulfate-reducing bacteria. The de-energization of the electrons received by $DvHc_3$ (increase in the reduction potential) leads to a decrease of the pK_a of the protonatable groups responsible for the two protons redox-Bohr effect (7.2 in oxidation stage 1 to 5.6 in stage *3*—Table 1.4), which can then be efficiently released in the periplasm. Therefore, the two electrons resulting from the oxidation of H_2 go through a transmembrane complex of electron transfer proteins to the cytoplasm where they can be used in sulfate respiration, whereas the two protons can then be used for ATP synthesis [145, 146].

The last example that illustrates the importance of the structural-functional relationship studies to understand the mechanistic features of MHC is illustrated by flavocytochrome c_3 from *S. frigidimarina*. Under anaerobic conditions and in the presence of fumarate as the sole terminal electron acceptor, *S. frigidimarina* produces large amounts of flavocytochrome that function as unidirectional terminal fumarate reductases in the periplasm [18] (see Sect. 1.2). Fumarate reductases catalyze the $2e^-/2H^+$ reduction of fumarate to succinate. The reducing equivalents used for this catalytic cycle are provided by lactate, formate or H_2, which are metabolic substrates commonly used by *Shewanella* spp. during anaerobic respiration.

The analysis of the heme oxidation profiles obtained for $Sffcc_3$ shows that the curves do not display the Nernst curve shape meaning that the electron affinity of the hemes is modulated during the protein oxidation (Fig. 1.12a). The clearest example of this effect is described by the oxidative curve of heme III, which shifts towards higher reduction potentials due to the significant redox interactions with heme II (65 mV). Consequently, in the first oxidation stage (loss of the first electron by the tetraheme domain) the reduction potential of heme III increases so that it remains essentially reduced. This can be further illustrated by the small contribution of the microstates with heme III (and IV) oxidized in the intermediate oxidation stages (Fig. 1.12b). Overall, these observations can be rationalized in functional terms, knowing that the semiquinone state of the FAD is not observed [147]. Oxidation of FAD leads to a sequential two-electron transfer from the heme domain to FAD when both hemes III and IV are thermodynamically biased to be reduced. Since the first electron is taken from heme IV by FAD, heme IV can readily be re-reduced by the neighboring heme III, thus facilitating the transfer of the second electron for reduction of FAD. Therefore, the strong redox interaction between hemes II and III, which increases the affinity of hemes III and IV for electrons in the catalytically relevant redox stages, favors the directionality of the intramolecular electron transfer from the heme domain to FAD. Furthermore, the concerted transfer of two electrons to FAD ensures that the lifetime of a semiquinone, and the possible consequence of producing reactive oxygen species, are minimized [147]. The data obtained for $Sffcc_3$ reveal how the thermodynamic properties of the individual hemes, modulated by the redox interactions with their closely spaced neighbors, can favor electron delivering to a two-electron active site by an array of single electron centers.

In summary, the studies revisited here illustrate how important is the fine tuning of the redox centers properties in MHC, a feature that cannot be accessed from simple macroscopic characterization.

1.7   Current Challenges and Perspectives

Although research on MHC has intensified considerably in the past decade, it still requires substantial inputs from various fields like engineering, genetic, microbiology, biochemistry, material sciences or electrochemistry. The several experimental and technological advances achieved in the last years opened new perspectives for the study of MHC. These include the development of efficient expression systems that allowed the expression of MHC containing up to twelve heme groups, as well as their cost-effective isotopic labeling. In parallel, the development of NMR equipment and technologies increased the spectral resolution and simplified the analysis of the experimental data, resulting in improved quality of three-dimensional structural models in the databases. The use of high-field NMR spectrometers equipped with cryoprobes favored the slow intermolecular electron exchange regime between different oxidation stages, a crucial step to identify the mechanistic functional features of MHC. Nowadays, the detailed structural and functional characterization of MHC is limited to small proteins (13 kDa MW) containing up to six interacting centers [148] or five centers in larger proteins (64 kDa MW) [147]. The recent methodological advances are expected to extend this characterization to MHC with higher number of redox centers and molecular weight. MHC containing more than six interacting centers have been shown to play crucial roles in extracellular electron transfer processes in DMR bacteria and in the performance of microbially catalyzed electrochemical systems, such as MFC, bioelectrochemical treatment (BET) or microbial electrosynthesis (MES). The power output (in MFC), the quantity of pollutant treated (in BET) or the nature of value added product synthesized (in MES) depend on the bacterial electron flux [149]. Effective transfer of electrons to electrodes is a crucial aspect in any bio-electrochemical system, and thus EET mechanisms are a main target for study and improvement. Additionally, many bacterial strains with specific electron transport pathways are unable to grow in a wide range of conditions requiring for the engineering of more efficient derivatives that can be used in bio-electrochemical systems of interest. Genetic engineering can be applied to modify the extracellular electron transport pathway or to deliver specific electron transport machinery into bacteria with increased growth rates. This has been successfully demonstrated by Jensen et al. [54] who introduced a synthetic electron conduit composed by the MtrA-MtrB-MtrC cytochrome complex from *S. oneidensis* into *E. coli* resulting in cells with increased rates for reduction of inorganic extracellular electron acceptors.

Altogether, the information collected by the application of these improved methodologies can then be used for the rational design of modified MHC, an essential step to develop exoelectrogens strains with increased respiratory rates.

Protein-protein interactions between redox partners is a topic that requires further analysis to clarify the respiratory networks in exoelectrogens microorganisms. Understanding the electron transfer pathways shall help engineer strains with enhanced performance for exoelectrogens-based biotechnological approaches of clear interest for immediate or near future applications.

References

1. Moore, G.R., Pettigrew, G.W.: Cytochromes c: Evolutionary, Structural and Physicochemical Aspects in Molecular Biology, 478 pp. Springer, Berlin (1990)
2. Bertini, I., et al.: Cytochrome c: occurrence and functions. Chem. Rev. **106**, 90–115 (2006)
3. Smith, L.J., et al.: Heme proteins-diversity in structural characteristics, function, and folding. Proteins **78**, 2349–2368 (2010)
4. Meyer, T.E., et al.: Occurrence and sequence of Sphaeroides Heme Protein and diheme cytochrome c in purple photosynthetic bacteria in the family *Rhodobacteraceae*. BMC Biochem. **11**, 24 (2010)
5. Leys, D., et al.: Crystal structures of an oxygen-binding cytochrome c from *Rhodobacter sphaeroides*. J. Biol. Chem. **275**, 16050–16056 (2000)
6. Stevens, J.M., et al.: Cytochrome c biogenesis System I. FEBS J. **278**, 4170–4178 (2011)
7. Simon, J., Hederstedt, L.: Composition and function of cytochrome c biogenesis System II. FEBS J. **278**, 4179–4188 (2011)
8. Allen, J.W.: Cytochrome c biogenesis in mitochondria-systems III and V. FEBS J. **278**, 4198–4216 (2011)
9. de Vitry, C.: Cytochrome c maturation system on the negative side of bioenergetic membranes: CCB or System IV. FEBS J. **278**, 4189–4197 (2011)
10. Mavridou, D.A., et al.: Probing heme delivery processes in cytochrome c biogenesis system I. Biochemistry **52**, 7262–7270 (2013)
11. Schulz, H., et al.: New insights into the role of CcmC, CcmD and CcmE in the haem delivery pathway during cytochrome c maturation by a complete mutational analysis of the conserved tryptophan-rich motif of CcmC. Mol. Microbiol. **37**, 1379–1388 (2000)
12. Feissner, R.E., et al.: ABC transporter-mediated release of a haem chaperone allows cytochrome c biogenesis. Mol. Microbiol. **61**, 219–231 (2006)
13. Sanders, C., et al.: The cytochrome c maturation components CcmF, CcmH, and CcmI form a membrane-integral multisubunit heme ligation complex. J. Biol. Chem. **283**, 29715–29722 (2008)
14. Verissimo, A.F., et al.: CcmI subunit of CcmFHI heme ligation complex functions as an apocytochrome c chaperone during c-type cytochrome maturation. J. Biol. Chem. **286**, 40452–40463 (2011)
15. Raven, E.: Heme Peroxidases. Royal Society of Chemistry, RSC Publishing, Cambridge (2015)
16. Gordon, E.H.J., et al.: Physiological function and regulation of flavocytochrome c_3, the soluble fumarate reductase from *Shewanella putrefaciens* NCIMB 400. Microbiology **144**, 937–945 (1998)
17. Pokkuluri, P.R., et al.: Structures and solution properties of two novel periplasmic sensor domains with c-type heme from chemotaxis proteins of *Geobacter sulfurreducens*: Implications for signal transduction. J. Mol. Biol. **377**, 1498–1517 (2008)
18. Turner, K.L., et al.: Redox properties of flavocytochrome c_3 from *Shewanella frigidimarina* NCIMB400. Biochemistry **38**, 3302–3309 (1999)
19. Leys, D., et al.: Structure and mechanism of the flavocytochrome c fumarate reductase of *Shewanella putrefaciens* MR-1. Nat. Struct. Biol. **6**, 1113–1117 (1999)

20. Taylor, P., et al.: Structural and mechanistic mapping of a unique fumarate reductase. Nat. Struct. Biol. **6**, 1108–1112 (1999)
21. Xavier, A.V.: Thermodynamic and choreographic constraints for energy transduction by cytochrome c oxidase. Biochim. Biophys. Acta **1658**, 23–30 (2004)
22. Esteve-Núñez, A., et al.: Fluorescent properties of c-type cytochromes reveal their potential role as an extracytoplasmic electron sink in *Geobacter sulfurreducens*. Environ. Microbiol. **10**, 497–505 (2008)
23. Breuer, M., et al.: Multi-haem cytochromes in *Shewanella oneidensis* MR-1: structures, functions and opportunities. J. R. Soc. Interface **12**, 102 (2015)
24. Page, C.C., et al.: Natural engineering principles of electron tunnelling in biological oxidation-reduction. Nature **402**, 47–52 (1999)
25. Pokkuluri, P.R., et al.: Structure of a novel dodecaheme cytochrome c from *Geobacter sulfurreducens* reveals an extended 12 nm protein with interacting hemes. J. Struct. Biol. **174**, 223–233 (2011)
26. Pokkuluri, P.R., et al.: Structure of a novel c_7-type three-heme cytochrome domain from a multidomain cytochrome c polymer. Protein Sci. **13**, 1684–1692 (2004)
27. Das, D.K., Medhi, O.K.: The role of heme propionate in controlling the redox potential of heme: square wave voltammetry of protoporphyrinato IX iron (III) in aqueous surfactant micelles. J. Inorg. Biochem. **70**, 83–90 (1998)
28. Maes, E.M., et al.: Ultrahigh resolution structures of nitrophorin 4: heme distortion in ferrous CO and NO complexes. Biochemistry **44**, 12690–12699 (2005)
29. Ma, J.G., et al.: The structural origin of nonplanar heme distortions in tetraheme ferricytochromes c_3. Biochemistry **37**, 12431–12442 (1998)
30. Tezcan, F.A., et al.: Effects of ligation and folding on reduction potentials of heme proteins. J. Am. Chem. Soc. **120**, 13383–13388 (1998)
31. Liu, Y., et al.: Replacement of the proximal histidine iron ligand by a cysteine or tyrosine converts heme oxygenase to an oxidase. Biochemistry **38**, 3733–3743 (1999)
32. Reddi, A.R., et al.: Thermodynamic investigation into the mechanisms of proton-coupled electron transfer events in heme protein maquettes. Biochemistry **46**, 291–305 (2007)
33. de Lacroix de Lavalette, A., et al.: Is the redox state of the c_i heme of the cytochrome b_6f complex dependent on the occupation and structure of the Q_i site and vice versa? J. Biol. Chem. **284**, 20822–20829 (2009)
34. Armstrong, F.A.: Evaluations of reduction potential data in relation to coupling, kinetics and function. J. Biol. Inorg. Chem. **2**, 139–142 (1997)
35. Bertini, I., et al.: Are unit charges always negligible? J. Biol. Inorg. Chem. **2**, 114–118 (1997)
36. Gunner, M.R., et al.: The importance of the protein in controlling the electrochemistry of heme metalloproteins: Methods of calculation and analysis. J. Biol. Inorg. Chem. **2**, 126–134 (1997)
37. Mauk, A.G., Moore, G.R.: Control of metalloprotein redox potentials: What does site-directed mutagenesis of hemoproteins tell us? J. Biol. Inorg. Chem. **2**, 119–125 (1997)
38. NaraySzabo, G.: Electrostatic modulation of electron transfer in the active site of heme peroxidases. J. Biol. Inorg. Chem. **2**, 135–138 (1997)
39. Warshel, A., et al.: Microscopic and semimacroscopic redox calculations: what can and cannot be learned from continuum models. J. Biol. Inorg. Chem. **2**, 143–152 (1997)
40. Zhou, H.X.: Control of reduction potential by protein matrix: lesson from a spherical protein model. J. Biol. Inorg. Chem. **2**, 109–113 (1997)
41. Sharma, S., et al.: A systematic investigation of multiheme c-type cytochromes in prokaryotes. J. Biol. Inorg. Chem. **15**, 559–571 (2010)
42. Banci, L., et al.: Mitochondrial cytochromes c: a comparative analysis. J. Biol. Inorg. Chem. **4**, 824–837 (1999)
43. Pettigrew, G.W., Moore, G.R.: Cytochromes c: biological aspects in molecular biology, 282 pp. Springer, Berlin (1987)
44. Nealson, K.H., Saffarini, D.: Iron and manganese in anaerobic respiration: environmental significance, physiology, and regulation. Annu. Rev. Microbiol. **48**, 311–343 (1994)

45. Lovley, D.R.: Dissimilatory metal reduction. Annu. Rev. Microbiol. **47**, 263–290 (1993)
46. Kappler, A., Straub, K.L.: Geomicrobiological cycling of iron. Rev. Mineral. Geochem. **59**, 85–108 (2005)
47. Hernandez, M.E., Newman, D.K.: Extracellular electron transfer. Cell. Mol. Life Sci. **58**, 1562–1571 (2001)
48. Erable, B., et al.: Application of electro-active biofilms. Biofouling **26**, 57–71 (2010)
49. Kumar, R., et al.: Exoelectrogens in microbial fuel cells toward bioelectricity generation: a review. Int. J. Energ. Res. **39**, 1048–1067 (2015)
50. Logan, B.E., et al.: Microbial fuel cells: methodology and technology. Environ. Sci. Technol. **40**, 5181–5192 (2006)
51. Lovley, D.R.: Electromicrobiology. Annu. Rev. Microbiol. **66**, 391–409 (2012)
52. Franks, A.E., et al.: Bacterial biofilms: the powerhouse of a microbial fuel cell. Biofuels **1**, 589–604 (2010)
53. Ren, H., et al.: A miniaturized microbial fuel cell with three-dimensional graphene macroporous scaffold anode demonstrating a record power density of over 10000 W m(-3). Nanoscale **8**, 3539–3547 (2016)
54. Jensen, H.M., et al.: Engineering of a synthetic electron conduit in living cells. Proc. Natl. Acad. Sci. U.S.A. **107**, 19213–19218 (2010)
55. Yi, H., et al.: Selection of a variant of *Geobacter sulfurreducens* with enhanced capacity for current production in microbial fuel cells. Biosens. Bioelectron. **24**, 3498–3503 (2009)
56. Methé, B.A., et al.: Genome of *Geobacter sulfurreducens*: metal reduction in subsurface environments. Science **302**, 1967–1969 (2003)
57. Heidelberg, J.F., et al.: Genome sequence of the dissimilatory metal ion-reducing bacterium *Shewanella oneidensis*. Nat. Biotechnol. **20**, 1118–1123 (2002)
58. Lovley, D.R., et al.: *Geobacter*: the microbe electric's physiology, ecology, and practical applications. Adv. Microb. Physiol. **59**, 1–100 (2011)
59. Aklujkar, M., et al.: Proteins involved in electron transfer to Fe(III) and Mn(IV) oxides by *Geobacter sulfurreducens* and *Geobacter uraniireducens*. Microbiology **159**, 515–535 (2013)
60. Orellana, R., et al.: Proteome of *Geobacter sulfurreducens* in the presence of U(VI). Microbiology **160**, 2607–2617 (2014)
61. Lovley, D.R.: Bug juice: harvesting electricity with microorganisms. Nat. Rev. Microbiol. **4**, 497–508 (2006)
62. Lovley, D.R.: Powering microbes with electricity: direct electron transfer from electrodes to microbes. Environ. Microbiol. Rep. **3**, 27–35 (2011)
63. Grisshammer, R., et al.: Expression in *Escherichia coli* of c-type cytochrome genes from *Rhodopseudomonas viridis*. Biochim. Biophys. Acta **1088**, 183–190 (1991)
64. Pollock, W.B., et al.: Expression of the gene encoding cytochrome c_3 from *Desulfovibrio vulgaris* (Hildenborough) in *Escherichia coli*: export and processing of the apoprotein. J. Gen. Microbiol. **135**, 2319–2328 (1989)
65. Thony-Meyer, L., et al.: *Escherichia coli* genes required for cytochrome c maturation. J. Bacteriol. **177**, 4321–4326 (1995)
66. Thony-Meyer, L.: Biogenesis of respiratory cytochromes in bacteria. Microbiol. Mol. Biol. Rev. **61**, 337–376 (1997)
67. Page, M.D., et al.: Contrasting routes of c-type cytochrome assembly in mitochondria, chloroplasts and bacteria. Trends Biochem. Sci. **23**, 103–108 (1998)
68. Arslan, E., et al.: Overproduction of the *Bradyrhizobium japonicum* c-type cytochrome subunits of the cbb_3 oxidase in *Escherichia coli*. Biochem Biophys Res Commun **251**, 744–747 (1998)
69. da Costa, P.N., et al.: Expression of a *Desulfovibrio* tetraheme cytochrome c in *Escherichia coli*. Biochem. Biophys. Res. Commun. **268**, 688–691 (2000)
70. Herbaud, M.L., et al.: *Escherichia coli* is able to produce heterologous tetraheme cytochrome c_3 when the *ccm* genes are co-expressed. Biochim. Biophys. Acta **1481**, 18–24 (2000)

71. Sanders, C., Lill, H.: Expression of prokaryotic and eukaryotic cytochromes *c* in *Escherichia coli*. Biochim. Biophys. Acta **1459**, 131–138 (2000)
72. Londer, Y.Y., et al.: Production and preliminary characterization of a recombinant triheme cytochrome c_7 from *Geobacter sulfurreducens* in *Escherichia coli*. Biochim. Biophys. Acta **1554**, 202–211 (2002)
73. Shi, L., et al.: Overexpression of multi-heme *c*-type cytochromes. Biotechniques **38**, 297–299 (2005)
74. Ozawa, K., et al.: A simple, rapid, and highly efficient gene expression system for multiheme cytochromes *c*. Biosci. Biotechnol. Biochem. **65**, 185–189 (2001)
75. Ozawa, K., et al.: Expression of a tetraheme protein, *Desulfovibrio vulgaris* Miyazaki F cytochrome c_3, in *Shewanella oneidensis* MR-1. Appl. Environ. Microbiol. **66**, 4168–4171 (2000)
76. Hau, H.H., Gralnick, J.A.: Ecology and biotechnology of the genus *Shewanella*. Annu. Rev. Microbiol. **61**, 237–258 (2007)
77. Jiang, X., et al.: Probing electron transfer mechanisms in *Shewanella oneidensis* MR-1 using a nanoelectrode platform and single-cell imaging. Proc. Natl. Acad. Sci. U.S.A. **107**, 16806–16810 (2010)
78. Brutinel, E.D., Gralnick, J.A.: Shuttling happens: soluble flavin mediators of extracellular electron transfer in *Shewanella*. Appl. Microbiol. Biotechnol. **93**, 41–48 (2012)
79. Gon, S., et al.: An unsuspected autoregulatory pathway involving apocytochrome TorC and sensor TorS in *Escherichia coli*. Proc. Natl. Acad. Sci. U.S.A. **98**, 11615–11620 (2001)
80. Schwalb, C., et al.: The tetraheme cytochrome CymA is required for anaerobic respiration with dimethyl sulfoxide and nitrite in *Shewanella oneidensis*. Biochemistry **42**, 9491–9497 (2003)
81. Marritt, S.J., et al.: The roles of CymA in support of the respiratory flexibility of *Shewanella oneidensis* MR-1. Biochem. Soc. Trans. **40**, 1217–1221 (2012)
82. Coursolle, D., Gralnick, J.A.: Reconstruction of extracellular respiratory pathways for iron (III) reduction in *Shewanella oneidensis* strain MR-1. Front. Microbiol. **3**, 56 (2012)
83. Hartshorne, R.S., et al.: Characterization of an electron conduit between bacteria and the extracellular environment. Proc. Natl. Acad. Sci. U.S.A. **106**, 22169–22174 (2009)
84. Richardson, D.J., et al.: The 'porin-cytochrome' model for microbe-to-mineral electron transfer. Mol. Microbiol. **85**, 201–212 (2012)
85. Fredrickson, J.K., et al.: Towards environmental systems biology of *Shewanella*. Nat. Rev. Microbiol. **6**, 592–603 (2008)
86. Fonseca, B.M., et al.: Mind the gap: cytochrome interactions reveal electron pathways across the periplasm of *Shewanella oneidensis* MR-1. Biochem. J. **449**, 101–108 (2013)
87. Zhang, H., et al.: *In vivo* identification of the outer membrane protein OmcA-MtrC interaction network in *Shewanella oneidensis* MR-1 cells using novel hydrophobic chemical cross-linkers. J. Proteome Res. **7**, 1712–1720 (2008)
88. White, G.F., et al.: Mechanisms of bacterial extracellular electron exchange. Adv. Microb. Physiol. **68**, 87–138 (2016)
89. Coursolle, D., et al.: The Mtr respiratory pathway is essential for reducing flavins and electrodes in *Shewanella oneidensis*. J. Bacteriol. **192**, 467–474 (2010)
90. Myers, J.M., Myers, C.R.: Isolation and sequence of *omcA*, a gene encoding a decaheme outer membrane cytochrome *c* of *Shewanella putrefaciens* MR-1, and detection of *omcA* homologs in other strains of *S. putrefaciens*. Biochim. Biophys. Acta **1373**, 237–251 (1998)
91. Myers, C.R., Nealson, K.H.: Bacterial manganese reduction and growth with manganese oxide as the sole electron-acceptor. Science **240**, 1319–1321 (1988)
92. Beliaev, A.S., et al.: MtrC, an outer membrane decahaem *c* cytochrome required for metal reduction in *Shewanella putrefaciens* MR-1. Mol. Microbiol. **39**, 722–730 (2001)
93. Alves, M.N., et al.: Characterization of the periplasmic redox network that sustains the versatile anaerobic metabolism of *Shewanella oneidensis* MR-1. Front. Microbiol. **6**, 665 (2015)

94. Gralnick, J.A., et al.: Extracellular respiration of dimethyl sulfoxide by *Shewanella oneidensis* strain MR-1. Proc. Natl. Acad. Sci. U.S.A. **103**, 4669–4674 (2006)
95. Lovley, D.R., et al.: Anaerobic production of magnetite by a dissimilatory iron-reducing microorganism. Nature **330**, 252–254 (1987)
96. Aklujkar, M., et al.: The genome sequence of *Geobacter metallireducens*: features of metabolism, physiology and regulation common and dissimilar to *Geobacter sulfurreducens*. BMC Microbiol. **9**, 109 (2009)
97. Caccavo Jr., F., et al.: *Geobacter sulfurreducens* sp. nov., a hydrogen- and acetate-oxidizing dissimilatory metal-reducing microorganism. Appl. Environ. Microbiol. **60**, 3752–3759 (1994)
98. Lin, W.C., et al.: *Geobacter sulfurreducens* can grow with oxygen as a terminal electron acceptor. Appl Environ Microb **70**, 2525–2528 (2004)
99. Levar, C.E., et al.: An inner membrane cytochrome required only for reduction of high redox potential extracellular electron acceptors. MBio **5**, e02034 (2014)
100. Zacharoff, L., et al.: Reduction of low potential electron acceptors requires the CbcL inner membrane cytochrome of *Geobacter sulfurreducens*. Bioelectrochemistry **107**, 7–13 (2016)
101. Shelobolina, E.S., et al.: Importance of *c*-type cytochromes for U(VI) reduction by *Geobacter sulfurreducens*. BMC Microbiol. **7**, 16 (2007)
102. Kim, B.C., Lovley, D.R.: Investigation of direct vs. indirect involvement of the *c*-type cytochrome MacA in Fe(III) reduction by *Geobacter sulfurreducens*. FEMS Microbiol. Lett. **286**, 39–44 (2008)
103. Ding, Y.H., et al.: Proteome of *Geobacter sulfurreducens* grown with Fe(III) oxide or Fe(III) citrate as the electron acceptor. Biochim. Biophys. Acta **1784**, 1935–1941 (2008)
104. Morgado, L., et al.: Thermodynamic characterization of a triheme cytochrome family from *Geobacter sulfurreducens* reveals mechanistic and functional diversity. Biophys. J. **99**, 293–301 (2010)
105. Lloyd, J.R., et al.: Biochemical and genetic characterization of PpcA, a periplasmic *c*-type cytochrome in *Geobacter sulfurreducens*. Biochem. J. **369**, 153–161 (2003)
106. Ding, Y.H., et al.: The proteome of dissimilatory metal-reducing microorganism *Geobacter sulfurreducens* under various growth conditions. Biochim. Biophys. Acta **1764**, 1198–1206 (2006)
107. Liu, Y., et al.: A trans-outer membrane porin-cytochrome protein complex for extracellular electron transfer by *Geobacter sulfurreducens* PCA. Environ. Microbiol. Rep. **6**, 776–785 (2014)
108. Liu, Y., et al.: Direct involvement of *ombB*, *omaB*, and *omcB* genes in extracellular reduction of Fe(III) by *Geobacter sulfurreducens* PCA. Front. Microbiol. **6**, 1075 (2015)
109. Pokkuluri, P.R., et al.: Outer membrane cytochrome *c*, OmcF, from *Geobacter sulfurreducens*: high structural similarity to an algal cytochrome c_6. Proteins **74**, 266–270 (2009)
110. Kim, B.C., et al.: OmcF, a putative *c*-type monoheme outer membrane cytochrome required for the expression of other outer membrane cytochromes in *Geobacter sulfurreducens*. J. Bacteriol. **187**, 4505–4513 (2005)
111. Kim, B.C., et al.: Insights into genes involved in electricity generation in *Geobacter sulfurreducens* via whole genome microarray analysis of the OmcF-deficient mutant. Bioelectrochemistry **73**, 70–75 (2008)
112. Inoue, K., et al.: Specific localization of the *c*-type cytochrome OmcZ at the anode surface in current-producing biofilms of *Geobacter sulfurreducens*. Env. Microbiol. Rep. **3**, 211–217 (2011)
113. Leang, C., et al.: Alignment of the *c*-type cytochrome OmcS along pili of *Geobacter sulfurreducens*. Appl. Environ. Microbiol. **76**, 4080–4084 (2010)
114. Nevin, K.P., et al.: Anode biofilm transcriptomics reveals outer surface components essential for high density current production in *Geobacter sulfurreducens* fuel cells. PLoS ONE **4**, e5628 (2009)

115. Mehta, T., et al.: Outer membrane c-type cytochromes required for Fe(III) and Mn(IV) oxide reduction in *Geobacter sulfurreducens*. Appl. Environ. Microbiol. **71**, 8634–8641 (2005)

116. Inoue, K., et al.: Purification and characterization of OmcZ, an outer-surface, octaheme c-type cytochrome essential for optimal current production by *Geobacter sulfurreducens*. Appl. Environ. Microbiol. **76**, 3999–4007 (2010)

117. Qian, X., et al.: Biochemical characterization of purified OmcS, a c-type cytochrome required for insoluble Fe(III) reduction in *Geobacter sulfurreducens*. Biochim. Biophys. Acta **1807**, 404–412 (2011)

118. Paquete, C.M., Louro, R.O.: Molecular details of multielectron transfer: the case of multiheme cytochromes from metal respiring organisms. Dalton Trans. **39**, 4259–4266 (2010)

119. Paixão, V.B., et al.: Redox linked conformational changes in cytochrome c_3 from *Desulfovibrio desulfuricans* ATCC 27774. Biochemistry **49**, 9620–9629 (2010)

120. Morgado, L., et al.: One simple step in the identification of the cofactors signals, one giant leap for the solution structure determination of multiheme proteins. Biochem. Biophys. Res. Commun. **393**, 466–470 (2010)

121. Londer, Y.Y., et al.: Heterologous expression of hexaheme fragments of a multidomain cytochrome from *Geobacter sulfurreducens* representing a novel class of cytochromes c. Protein Expr. Purif. **39**, 254–260 (2005)

122. Londer, Y.Y., et al.: Heterologous expression of dodecaheme "nanowire" cytochromes c from *Geobacter sulfurreducens*. Protein Expr. Purif. **47**, 241–248 (2006)

123. Fernandes, A.P., et al.: Isotopic labeling of c-type multiheme cytochromes overexpressed in *E. coli*. Protein Expr. Purif. **59**, 182–188 (2008)

124. Fonseca, B.M., et al.: Efficient and selective isotopic labeling of hemes to facilitate the study of multiheme proteins. Biotechniques **52**, 4 (2012)

125. Morgado, L., et al.: Backbone, side chain and heme resonance assignments of the triheme cytochrome PpcA from *Geobacter sulfurreducens*. Biomol. NMR Assign. **5**, 113–116 (2011)

126. Dantas, J.M., et al.: Evidence for interaction between the triheme cytochrome PpcA from *Geobacter sulfurreducens* and anthrahydroquinone-2,6-disulfonate, an analog of the redox active components of humic substances. Biochim. Biophys. Acta **1837**, 750–760 (2014)

127. Dantas, J.M., et al.: Backbone, side chain and heme resonance assignments of the triheme cytochrome PpcD from *Geobacter sulfurreducens*. Biomol. NMR Assign. **9**, 211–214 (2015)

128. Morgado, L., et al.: Revealing the structural origin of the redox-Bohr effect: the first solution structure of a cytochrome from *Geobacter sulfurreducens*. Biochem. J. **441**, 179–187 (2012)

129. Pokkuluri, P.R., et al.: Family of cytochrome c_7-type proteins from *Geobacter sulfurreducens*: structure of one cytochrome c_7 at 1.45 Å resolution. Biochemistry **43**, 849–859 (2004)

130. Morgado, L., et al.: Structural insights into the modulation of the redox properties of two *Geobacter sulfurreducens* homologous triheme cytochromes. Biochim. Biophys. Acta **1777**, 1157–1165 (2008)

131. Pokkuluri, P.R., et al.: Structural characterization of a family of cytochromes c_7 involved in Fe(III) respiration by *Geobacter sulfurreducens*. Biochim. Biophys. Acta **1797**, 222–232 (2010)

132. Pessanha, M., et al.: Redox characterization of *Geobacter sulfurreducens* cytochrome c_7: physiological relevance of the conserved residue F15 probed by site-specific mutagenesis. Biochemistry **43**, 9909–9917 (2004)

133. Seidel, J., et al.: MacA is a second cytochrome c peroxidase of *Geobacter sulfurreducens*. Biochemistry **51**, 2747–2756 (2012)

134. Santos, T.C., et al.: Diving into the redox properties of *Geobacter sulfurreducens* cytochromes: a model for extracellular electron transfer. Dalton Trans. **44**, 9335–9344 (2015)

135. Salgueiro, C.A.: The multifaceted role of cytochromes c_7 in metal reduction—a structural and functional overview. Global J. Biochem. **3**, 5 (2012)
136. Catarino, T., Turner, D.L.: Thermodynamic control of electron transfer rates in multicentre redox proteins. ChemBioChem **2**, 416–424 (2001)
137. Paquete, C.M., Louro, R.O.: Unveiling the details of electron transfer in multicenter redox proteins. Acc. Chem. Res. **47**, 56–65 (2014)
138. Turner, D.L., et al.: NMR studies of cooperativity in the tetrahaem cytochrome c_3 from *Desulfovibrio vulgaris*. Eur. J. Biochem. **241**, 723–731 (1996)
139. Salgueiro, C.A., et al.: Assignment of the redox potentials to the 4 hemes in *Desulfovibrio vulgaris* cytochrome c_3 by 2D-NMR. FEBS Lett. **314**, 155–158 (1992)
140. Santos, H., et al.: NMR studies of electron transfer mechanisms in a protein with interacting redox centres: *Desulfovibrio gigas* cytochrome c_3. Eur. J. Biochem. **141**, 283–296 (1984)
141. Badziong, W., et al.: Isolation and characterization of *Desulfovibrio* growing on hydrogen plus sulfate as sole energy-source. Arch. Microbiol. **116**, 41–49 (1978)
142. Xavier, A.V.: Energy transduction coupling mechanisms in multiredox center proteins. J. Inorg. Biochem. **28**, 239–243 (1986)
143. LeGall, J., Fauque, G.: Dissimilatory Reduction of Sulfur Compounds in Biology of Anaerobic Microorganisms, pp. 587–639. Wiley, New York (1988)
144. Louro, R.O., et al.: Redox-Bohr effect in the tetrahaem cytochrome c_3 from *Desulfovibrio vulgaris*: a model for energy transduction mechanisms. J. Biol. Inorg. Chem. **1**, 34–38 (1996)
145. Chen, L., et al.: Isolation and characterization of flavoredoxin, a new flavoprotein that permits *in vitro* reconstitution of an electron transfer chain from molecular hydrogen to sulfite reduction in the bacterium *Desulfovibrio gigas*. Arch. Biochem. Biophys. **303**, 44–50 (1993)
146. Rossi, M., et al.: The *hmc* operon of *Desulfovibrio vulgaris* subsp. *vulgaris* Hildenborough encodes a potential transmembrane redox protein complex. J. Bacteriol. **175**, 4699–4711 (1993)
147. Pessanha, M., et al.: Tuning of functional heme reduction potentials in *Shewanella* fumarate reductases. Biochim. Biophys. Acta **1787**, 113–120 (2009)
148. Louro, R.O., et al.: Cooperativity between electrons and protons in a monomeric cytochrome c_3: the importance of mechano-chemical coupling for energy transduction. ChemBioChem **2**, 831–837 (2001)
149. Mohan, S.V., et al.: Biotreatability studies of pharmaceutical wastewater using an anaerobic suspended film contact reactor. Water Sci. Technol. **43**, 271–276 (2001)
150. Dantas, J.M., et al.: Molecular interaction studies revealed the bifunctional behavior of triheme cytochrome PpcA from *Geobacter sulfurreducens* toward the redox active analog of humic substances. Biochim. Biophys. Acta **1847**, 1129–1138 (2015)
151. Paquete, C.M., et al.: Thermodynamic and kinetic characterisation of individual haems in multicentre cytochromes c_3. Biochim. Biophys. Acta **1767**, 1169–1179 (2007)
152. Moss, G.P.: Nomenclature of tetrapyrroles. Recommendations 1986 IUPAC-IUB Joint Commission on Biochemical Nomenclature (JCBN). Eur. J. Biochem. **178**, 277–328 (1988)
153. Dantas, J.M., et al.: The structure of PccH from *Geobacter sulfurreducens*—a novel low reduction potential monoheme cytochrome essential for accepting electrons from an electrode. FEBS J. **282**, 2215–2231 (2015)
154. Simões, P. et al.: Refinement of the three-dimensional structures of cytochrome c_3, from *Desulfovibrio vulgaris* Hildenborough at 1.67 angstrom resolution and from *Desulfovibrio desulfuricans* ATCC 27774 at 1.6 angstrom resolution. Inorg. Chim. Acta **273**, 213–224 (1998)
155. Jensen, L.M., et al.: In crystallo posttranslational modification within a MauG/pre-methylamine dehydrogenase complex. Science **327**, 1392–1394 (2010)
156. DeLano, W.L.: The PyMOL molecular graphics system. http://www.pymol.org (2002)
157. Leys, D., et al.: Crystal structures at atomic resolution reveal the novel concept of "electron-harvesting" as a role for the small tetraheme cytochrome *c*. J. Biol. Chem. **277**, 35703–35711 (2002)

158. Clarke, T.A., et al.: Structure of a bacterial cell surface decaheme electron conduit. Proc. Natl. Acad. Sci. U.S.A. **108**, 9384–9389 (2011)
159. Edwards, M.J., et al.: The X-ray crystal structure of *Shewanella oneidensis* OmcA reveals new insight at the microbe-mineral interface. FEBS Lett. **588**, 1886–1890 (2014)